Communicating in Geography and the Environmental Sciences

Communicating in Geography and the Environmental Sciences

Iain Hay & Philip Giles

Second Canadian Edition

OXFORD
UNIVERSITY PRESS

Oxford University Press is a department of the University of Oxford.
It furthers the University's objective of excellence in research, scholarship,
and education by publishing worldwide. Oxford is a registered trade mark of
Oxford University Press in the UK and in certain other countries.

Published in Canada by
Oxford University Press
8 Sampson Mews, Suite 204,
Don Mills, Ontario M3C 0H5 Canada

www.oupcanada.com

First Canadian Edition published in 2011

Second Canadian Edition published in 2015

Communicating in Geography and the Environmental Sciences, Third Edition, was originally published in English in 2006 by Oxford University Press Australia & New Zealand, 253 Normandy Road, South Melbourne, Victoria, 3205 Australia with the ISBN 9780195517613. This adapted edition is published by arrangement. *Oxford University Press Canada is solely responsible for this adaptation from the original work*. Copyright © 2006, 2002, 1996 Iain Hay.

Library and Archives Canada Cataloguing in Publication

Hay, Iain, 1960–, author, Communicating in geography and the environmental sciences /
Iain Hay & Philip Giles. — Second Canadian edition.

Includes index.ISBN 978-0-19-900741-7 (pbk.)

1. Communication in geography. I. Giles, Philip (Professor of geography), author II. Title.

G70.H39 2014 910.72 C2014-905305-3

Cover images (clockwise from top left): ©iStockPhoto.com/BartCo,
©iStockPhoto.com/R-J-Seymour, ©iStockPhoto.com/Nikada, ©iStockPhoto.com/kadmy

Oxford University Press is committed to our environment.
This book is printed on Forest Stewardship Council® certified paper,
and contains 100% post-consumer waste.

Printed and bound in Canada

1 2 3 4 — 18 17 16 15

Contents

Boxes, Figures, and Tables x
Acknowledgements xiii
Introductory Comments for All Readers xiv

1 Introducing Writing in Geography and the Environmental Science 1

Key Topics 1
Why Write? 2
Keys to Good Writing 3
  Write for your audience 4
  Avoid plagiarism 5
  Get help with writing 5
How to Improve the Quality of Your Written Expression 6
  Write fluently and succinctly 6
  Pay attention to paragraph structure 6
  Use correct grammar, punctuation, and spelling 7
  Keep your assignment to a reasonable length 8
How to Improve the Presentation of Your Work 8
  Select a brief and informative title 8
  Make sure your work is legible and well presented 9
  Use headings to show structure 10
  Format source citations and references properly 11
  Demonstrate a level of individual scholarship 13
How to Use Supplementary Materials Correctly 14
  Make effective use of figures and tables 15
  Present the supplementary materials correctly 15
Further Reading 17

2 Finding, Evaluating, and Using Sources 18

Key Topics 18
Why Are Sources Important in Scholarly Writing? 19
Differentiating Academic from Non-Academic Sources 20
Finding Sources with Useful and Relevant Information 20
  Ensure you have enough sources 22
  Types of sources 23
Evaluating the Credibility of Sources 26
  Questions to help you evaluate the credibility of sources 27
Incorporating Reference Material into Your Writing 33
  Acknowledge your sources 33
  Paraphrase and quote your source material 35
  Common knowledge does not need to be cited 38
Steering Clear of Academic Dishonesty, Including Plagiarism 39
Preparing a Research Ethics Application or Environmental Impact Statement 41
Further Reading 45

3 Writing an Essay 46

Key Topics 46
Quality of Argument 48
- Ensure that the essay addresses the question fully 48
- Be sure that your essay is developed logically 49
- Write a thesis statement 49
- Create a structure for your essay 51
- Review and revise your essay drafts 54

Quality of evidence 57
- Ensure your essay is well supported by evidence and examples 57

A Self-Assessment Form for Your Essay 58
Further Reading 60

4 Writing a Research Report 62

Key Topics 62
Why Write a Research Report? 62
What Are Readers Looking For in a Report? 63
Writing a Research Report 65
- Preliminary material 67
- Introduction 71
- Literature review 72
- Materials and methods 74
- Results 76
- Discussion and conclusion 77
- Recommendations 78
- Reference list 78
- Appendices 79
- Written expression and presentation 79

Writing a Laboratory Report 80
Writing a Research Proposal 84
- Purpose 84
- Characteristics 84

Further Reading 86

5 Writing an Annotated Bibliography, Summary, or Review 88

Key Topics 88
Preparing an Annotated Bibliography 88
- What is the purpose of an annotated bibliography? 90
- What is the reader looking for in an annotated bibliography? 90

Writing a Summary or Précis 91
- What is the reader of a summary looking for? 92

Writing a Review 95
- Summary: What is this work about? 96
- Analysis: What are the strengths and weaknesses of this work? 101
- Evaluation: What is this work's contribution to the discipline? 104
- Written expression and presentation of the review 105

Examples of published reviews 105
Further Reading 107

6 Referencing and Language Matters 109

Key Topics 109
What Are Citations and References? Why Do We Need Them? 109
Principal Reference Systems: Author–Date and Note Systems 111
Variants of reference systems: Reference styles 112
The author–date (Harvard) system 114
The note system 127
Avoiding Biased Language and Stereotyping 134
Sexual bias and stereotyping 135
Other biases and stereotyping 135
Salient Notes on Punctuation and Spelling 136
Comma [,] 137
Period [.] 138
Ellipsis points [. . .] 139
Semicolon [;] 139
Colon [:] 139
Quotation marks (double or single) [" " or ' '] 140
Apostrophe ['] 140
Upper case letters 141
Numerals 141
Further Reading 142

7 Making a Poster 143

Key Topics 143
Why Make a Poster? 143
How Are Posters Produced? 144
What Are Your Markers Looking for in a Poster? 145
Guidelines for Designing a Poster 147
Preliminary layout 149
Organization 149
Text 151
Colour 153
Tables, figures, and photographs 154
List of sources 157
Further Reading 157

8 Communicating with Graphs and Tables 159

Key Topics 159
Why Communicate with Graphs and Tables? 159
General Guidelines for Communicating Clearly with Graphs 160
Computerized versus Manual Production of Graphs 161
Different Types of Graphs 163
Scatter plots 163
Line graphs 165
Bar graphs 169

Histograms 173
Population pyramids or age–sex pyramids 179
Pie graphs 181
Graphs with logarithmic axes 182
Tables 185
Elements of a table 186
Designing a table 187
Further Reading 189

9 Communicating with Maps 190

Key Topics 190
What Is the Purpose of a Map? 190
What Are the Different Types of Maps? 191
Dot maps 191
Proportional dot maps 192
Choropleth maps 195
Isoline maps 197
Other thematic maps 197
Topographic maps 199
Orthophoto maps 201
How Is Map Scale Related to Detail? 201
Cartographic scale 201
How to display the scale of map 202
Scale and map generalization 203
What Are the Characteristics of a Good Map? 204
Standard map elements 206
Map design elements 207
Further Reading 208

10 Preparing and Delivering an Oral Presentation 209

Key Topics 209
Why Is Public Speaking Important? 209
How to Prepare for an Oral Presentation 214
Establish the context and goals 214
Organize the material for presentation 215
Structure your presentation 216
Prepare your text and aids to delivery 219
Rehearse 224
Final points of preparation 226
How to Deliver a Successful Oral Presentation 227
How to Cope with Questions from the Audience 230
Further Reading 232

11 Writing for the Media and Public Audiences 234

Key Topics 234
Communicating with Public Audiences through the Media 234
Types of media 236

Using Media Releases to Disseminate Information 237
How to write a media release 239
How to format a media release 243
How to enhance a media release for the digital environment 246
How to follow up on a media release 246
Disseminating Information by Self-Publishing on the Internet 247
Opportunity comes with responsibility 247
Is self-publishing effective? 249
Content and presentation 249
Conclusion 250
Further Reading 250

12 Succeeding in Examinations 252

Key Topics 252
Purposes of Examinations 252
Types of Examinations 253
How to Prepare for an Examination 254
Ongoing preparatory activities during the term 254
Activities for successful exam preparation during review period 256
Preparations on the day of the examination 259
Techniques for Success in Examinations 260
Steps to take before answering your first question 260
Advice for questions requiring written answers 261
Advice for multiple-choice examinations 264
Advice for oral examinations 265
Advice for open-book examinations 267
Advice for take-home examinations 268
Advice for online examinations 269
Further Reading 269

Appendix 271
Glossary 279
References 285
Credits 291
Index 293

Boxes, Figures, and Tables

Boxes

1.1 Parts of a paragraph 7
1.2 Examples of hierarchical heading styles 12
1.3 Using supplementary materials 16
2.1 Distinguishing between the "Internet" and the "web" 22
2.2 The importance of formal peer review 24
2.3 Wikipedia: Facts or fiction? 29
2.4 Evaluating the credibility of web pages: An example 32
2.5 Open-access publishing 34
2.6 Reasons for acknowledging sources 35
2.7 Thinking about plagiarism 42
3.1 Types of essays in geography and the environmental sciences 47
3.2 What is a marker looking for in your essay? 47
3.3 Forming an essay structure 52
3.4 Guide to the introduction, body, and conclusion of an essay 53
3.5 Essay assessment form 58
4.1 Five investigative questions that characterize research reports 64
4.2 Examples of information abstracts 69
4.3 Examples of descriptive abstracts 70
4.4 The introduction of a report 72
4.5 Research report assessment form 81
5.1 Examples of items in annotated bibliographies 89
5.2 Key information in an annotated bibliography 90
5.3 Summary (or précis) assessment form 93
5.4 Assessment form for a review 97
5.5 Examples of reviews 105
6.1 Bibliographic-management software programs 110
6.2 Common in-text citation formatting errors 115
6.3 Common spelling "mistakes": Right spelling, wrong word 137
7.1 Principles of poster production 146
7.2 Poster assessment form 147
7.3 Suggested type sizes for posters 153
7.4 Characteristics of type 153
8.1 Essential information on graphs 162
9.1 Topological maps and cartograms 198
9.2 Scale and mental maps 204
9.3 Computerized versus manual map creation 205
10.1 Assessment form for an oral presentation 211
10.2 Some organizational frameworks for an oral presentation 216
10.3 Introducing an oral presentation 217
10.4 Effective use of presentation software 221

10.5 Visual aids: Alternatives to PowerPoint 223
10.6 Directions for lulling an audience to sleep 227
11.1 Why communicate with a public audience? 235
11.2 What are users of media releases looking for? 238
11.3 Targeting a media release 239
11.4 Suggested layout of a media release 244
11.5 Example of a media release to make an announcement, or to publicize an event or innovation 244
12.1 Questions to ask about your examination 256
12.2 Common questions in oral exams 267

Figures

7.1 Examples of different poster structures 151
7.2 Converting a numerical table into a poster-ready figure 156
8.1 Example of a scatter plot. *Total fertility rates vs life expectancy at birth, selected countries, 2005–10* 164
8.2 Example of a line graph. *Cumulative number of sites inscribed on the World Heritage List, 1977–2013* 165
8.3 Example of a line graph. *Trends in fine particulate matter ($PM_{2.5}$) levels in the air at selected locations in British Columbia, 1998–2008* 166
8.4 Example of a line graph. *Mean NDVI (Normalized Difference Vegetation Index)—a measure of vegetation "vigour"—following application of three different herbicides* 166
8.5 Example of a graph showing variation within data (mean and range). *Mean (including trend line) and range of NDVI (Normalized Difference Vegetation Index)—a measure of vegetation "vigour"—measurements* 167
8.6 Example of a graph showing variation within data (data range, mean, and standard deviation). *Variations about the mean (range and standard deviation) in NDVI (Normalized Difference Vegetation Index)—a measure of vegetation "vigour"—measurements* 168
8.7 Example of a line graph with two vertical axes. *Changes in lake chemistry, Clearwater Lake, Ontario, 1973–1999* 169
8.8 Example of a vertical bar graph. *Dwelling starts in Canada, seasonally adjusted data at annual rates, 2008* 170
8.9 Example of a horizontal bar graph. *Forest area harvested, by province, 2007* 170
8.10 Example of a bar graph. *Number of persons with asthma in Atlantic provinces, 2008* 171
8.11 Example of a subdivided bar graph. *Comparison of part-time and full-time employment in Canada, 2004 and 2008* 171
8.12 Example of a subdivided 100% bar graph. *Highest level of education attained in Atlantic Provinces, 2006 Census* 172
8.13 Example of a bar graph depicting positive and negative values. *Annual population growth rates for selected countries, 2000–5* 172

8.14 Examples of histograms based on the same data but different class intervals 175
8.15 Example of a population pyramid. *Age–sex structure of Mexico's population, 2010 (total population = 117 886 000)* 180
8.16 Example of a pie graph. *Greenhouse gas emissions by activity sector, Canada, 2005* 181
8.17 Comparison of data displayed on (a) arithmetic vertical axis and (b) logarithmic vertical axis 183
8.18 Example of a graph with two logarithmic axes (a "log-log" graph). *Relationship between GDP and population, selected countries, 2008* 184
9.1 Example of a dot map. *Swine farms in Southern Ontario, 2005* 192
9.2 Example of a proportional dot map. *Population of Canadian provinces and territories, 2008* 193
9.3 Example of a completed proportional circle scale 194
9.4 Example of a split proportional circle 194
9.5 Example of a choropleth map. *Migratory exchanges within Montreal Census Metropolitan Area, 2001–2006* 196
9.6 Example of an isoline map. *Mean annual precipitation in Mexico* 198
9.7 Example of a cartogram. *World reserves of oil, 2004* 200

Tables

1.1 Supplementary materials and how they are named 14
2.1 Some criteria for differentiating academic from non-academic sources 21
2.2 Evaluating web pages 31
4.1 Sections of a report 67
8.1 Types of graphs and their nature or function 163
8.2 Classes for Yarmouth mean daily temperatures, 2007, with 6 C° intervals starting at −18°C 178
8.3 Classes for Yarmouth mean daily temperatures, 2007, with 5 C° intervals starting at −15°C 178
8.4 Frequency table for data from Yarmouth mean daily temperatures, 2007 179
8.5 Example of a table. Land and water areas of Canadian provinces and territories 186
9.1 Some common isolines and the variables they depict 197
9.2 Forms of cartographic scale portrayed on maps 203
12.1 Types of examinations and their characteristics 253
12.2 Suggested time allocation for a three-hour examination 261

Acknowledgements

I am indebted to Iain Hay for writing *Communicating in Geography and the Environmental Sciences*, now in its fourth edition, from which the first Canadian edition was adapted in 2010. Iain's work laid a strong foundation, and one of my goals has been to retain his clear, direct style as much as possible. I would like to thank the reviewers for their comments, which helped to identify key areas for improvements and updates for this second Canadian edition. The guidance of Jodi Lewchuk, Senior Developmental Editor at Oxford University Press Canada, during the revision process is gratefully acknowledged. Other OUP Canada staff who deserve my recognition for their work on the first and second Canadian editions are Kate Skene, Caroline Starr, Meg Patterson, Leah-Ann Lymer, and Mark Thompson. Meticulous editing by Colleen Ste. Marie corrected my errors and greatly improved the manuscript. Finally I would like to thank my wife, Yvonne; my parents; and my colleagues at Saint Mary's University for their support.

Philip Giles

Introductory Comments for All Readers

Communicating in Geography and the Environmental Sciences, second Canadian edition, is about communicating effectively in academic settings. It discusses the character and practice of the most common forms of academic presentation skills used by students of geography and the environment. This book also addresses some less common forms of communication expected of post-secondary students. The 12 chapters outline the "whys" and "hows" of academic writing: essays; research and laboratory reports; reviews, summaries, and annotated bibliographies; sources; referencing and language matters; graphs and tables; maps; posters; oral presentations; media releases; and examinations. Information on the ways that these forms of presentation are commonly assessed is also an important element of the book.

Effective communication is a vital component of intellectual endeavour. Most audiences expect that certain conventions will be upheld or followed by people communicating to them through specific media. Unfortunately, however, many students do not know the accepted cues, clues, ceremonies, conventions, and characteristics associated with formal (academic) communication. In other words, some students do not know how to "make the grade." It is primarily because of that problem and because of the lack of specific, relevant advice to address it that this book was written. In the pages that follow we have tried to demystify the conventions of communication associated with post-secondary education.

The book serves a number of other important purposes, particularly the following:

- To help improve teaching, learning, and assessment: Many academics now find themselves asked to do more with less—to do more teaching, more research, more administration, and more community service without corresponding increases in support or resources. One means of coping with this tension is to teach more efficiently and effectively. This book is an attempt to contribute to those ends.
- To provide students with useful career skills: In the context of emerging patterns of work and work organization, the ability of university graduates to convey ideas and information is becoming increasingly important. Indeed, mention of this skill surfaces repeatedly in the reports of government and business think-tanks and academic authors. In some recognition of career considerations, this book makes an effort to contribute to the development of a range of communication skills.

The first part of the book is oriented to communication in the written form. Chapter 1 is an introduction to the practice of writing in geography and environmental sciences, while Chapter 2 is devoted to finding, evaluating, and using sources. Chapters 3–5 cover the writing of essays, reports, annotated bibliographies, summaries, and reviews. These are followed by a chapter on referencing and language matters (Chapter 6). The second part of the book, Chapters 7–12, explores other methods of communicating that a geography or environmental science student may be expected to use: posters, graphs and tables, maps, oral presentations, media releases, and examinations. In conjunction with the knowledge and skills learned in courses and from reading subject-specific material, this book is designed as a whole to help geography and environmental science students develop better communication skills.

Several of the chapters have a common framework comprising four main sections. First, there is an explanation of the specific type of communication being discussed. This explanation takes the form of an answer to a question such as "Why make a poster?" This is followed by a section containing a broad, conceptual statement of what markers are looking for in the particular type of communication under consideration. The third section consists of an outline of the means of making effective contact with one's audience. Where possible, this discussion is structured around an explicit statement of the sorts of criteria that most assessors have been found to use when marking student work. The fourth section is an example of an assessment form for the type of work being discussed. Professors may find these lists useful as marking guides, for example, to ensure that a broad range of matters is considered during marking and to offer consistency in marking practice; and they may be of considerable benefit to students seeking either a checklist that might be used in critical self-review or an indication of the sorts of things that are considered by markers.

A message to student readers

It is unlikely that you will have to use all of the material in this book in any one university course. Instead, this is a book written to be used throughout your entire degree program (and beyond). So, think of the book as comprising sections that will be important over several years, not limited to just one course.

When reading the table of contents, you may question why there are chapters on topics such as making posters, giving oral presentations, and writing for the media and public audiences. These topics are included with an eye to developing a broader set of communication skills beyond writing essays and reports. It is not uncommon for upper-level students to conduct research projects that they present to an audience in a poster or an oral presentation. It is less common for post-secondary students to write media releases for their degree programs, but you may participate in extracurricular activities for which you must communicate with the media or public audiences. Furthermore, while you may not encounter these exercises during your undergraduate program, you may go on to pursue a graduate degree or a career in geography or the environmental sciences. If so, there is a good chance you will need to develop proficiency in communicating in forms other than traditional academic writing.

An understanding of your audience's expectation *before* you undertake an assignment ought to help you prepare better work than might otherwise have been possible. You will find it helpful to review the relevant chapters and the assessment forms before you begin an assignment requiring some specific form of communication. You will then be able to undertake the assignment with an understanding of the relevant conventions of communication. When you have finished a draft of your work, try marking it yourself, using the checklist as a guide. If you have a patient and thoughtful friend, ask him or her to do this for you, too. The checklist will help to ensure that you and your friend give consideration to the broad range of issues likely to be examined by any marker, whether or not that person actually uses the forms when marking. If something in the assessment form does not make sense to you, consult the material in the appropriate chapter for an explanation. In this way you may be able to illuminate and correct any shortcomings in your work before they are uncovered by your professor. The end results of this process ought to be better communication and better grades.

As well as heeding the content of this book, you should read books, journal articles, and other people's essays critically; pay attention to the ways that both effective and poor speakers present themselves and their material; and be critical of graphs and maps. Evaluate what works and what does not. Learn from your observations and attempt to forge your own distinctive approach to communicating. The style you develop may be very effective and yet transgress some of the guidelines set out in the following pages. This should not be a matter of concern: your individuality and imagination are to be celebrated and encouraged, not condemned and excised.

A message to professors

Despite the revisions (and incorporation of the perspectives of two authors from different continents), this edition is still not a book of magic. We do not wish to make wild claims about the benefits that might flow from student and teacher use of the pages that follow. However, on the basis of experiences with the earlier editions, we do have reason to believe that if material from this book is referred to and incorporated into the teaching *and* marking within a discipline or throughout a degree program, there is likely to be an improvement in student communication skills. A number of simple and effective strategies for using this book have yielded positive results:

- Use the assessment forms provided here as a means of encouraging your students to think about their audience's expectations.
- Encourage students, by whatever means you consider appropriate, to read critically the assessment forms and explanatory notes in the relevant chapter of the book before they begin an assigned task.
- Apply strategies that encourage students to use marking forms to assess their own work before submitting it for peer review or final assessment.
- Use peer review methods as means of encouraging students to think critically about communication and the expectations of author and audience. The assessment forms found in most of the chapters in this book are useful preliminary frameworks for peer review.
- Use the assessment forms, and use them repeatedly, as the foundation for your own marking.

These strategies, singly and in combination, while requiring little extra teaching effort, offer the potential for improvements in written, oral, and graphic communication skills. Why not give them a try?

Comments about the second Canadian edition

The Canadian editions of *Communicating in Geography and the Environmental Sciences* are adapted from Iain Hay's book of the same title, now in its fourth edition, written primarily for audiences in Australia and New Zealand. The objectives of the first Canadian edition were (1) to adapt terminology and examples for a Canadian audience, (2) to apply some modifications to the book's structure, including the appearance of a new chapter on sources and rearranging the order of other chapters, and (3) to update and refresh the text, examples, and references.

For the second Canadian edition, the goal was to strengthen and update the book based on reflection and reviewers' comments. A weakness in the structure of the previous edition led to the decision to create a new chapter to introduce the practice of writing in geography and the environmental sciences.

Chapter 1 now contains material that previously was found mostly in the chapter on essay writing despite having relevance to other forms of written assignments. There remains a chapter on essay writing specifically, Chapter 3.

The chapter on finding, evaluating, and using sources (Chapter 2) has been moved from a position following the chapters discussing various writing assignments (Chapters 3–5) to one that precedes them. This reflects the fact that searching for information and sources is normally done *before* writing actually begins. The remainder of the chapters are found in the same order as in the first Canadian edition.

An important new feature of this book is a short sample paper in the Appendix. It is presented in two versions: (1) a "poor" version that contains examples of common errors and ineffective writing submitted by students; and (2) a corresponding "better" version. Both versions are annotated with comments. The sample paper should not be considered as a comprehensive review of all material presented about writing in the book; in particular, space restrictions limit its length. The objective is to show students an example of a poor paper riddled with common errors and then show how it can be revised into a better paper with attention to writing, citation, and reference matters. Keeping the sample paper short allows students to compare the two versions.

Comments received from reviewers of the previous edition were used to guide many of the revisions. There has been some material added (e.g., new sections about differentiating academic from non-academic sources, writing a research proposal, preparing a research ethics application, and effective use of presentation software), as well as some trimming conducted throughout the book. Finally, the entire text was reviewed and edits have been made to improve clarity, reduce redundancy, and update material where appropriate.

1 Introducing Writing in Geography and the Environmental Sciences

When something can be read without effort,
great effort has gone into its writing.

—Enrique Jardiel Poncela

Don't try to figure out what other people want to hear from you;
figure out what you have to say.

—Barbara Kingsolver

Key Topics

- Why write?
- Keys to good writing
- How to improve the quality of your written expression
- How to improve the presentation of your work
- How to use supplementary materials correctly

Our introductory comments present the case for the importance of effective communication and declare our objectives for helping students to develop their skills in the various forms of communication required in academic settings. In the first part of this book (Chapters 1–6) we focus on written communication. This chapter begins by briefly **arguing** the case for writing in general before going on to **discuss** how to develop and improve your writing and presentation skills. The use of supplementary materials to enhance the communication effectiveness of assignments is also **considered**. Chapter 2 is devoted to a discussion of finding, evaluating, and using sources. Several chapters are devoted to writing for the different types of assignments that geography and environmental science students are likely to encounter: essays (Chapter 3); reports (Chapter 4); and annotated bibliographies, summaries, and reviews (Chapter 5). Matters relating to referencing and language are presented in Chapter 6. (See the Glossary at the end of this book for a discussion of terms highlighted in bold throughout the text.)

Throughout this chapter and others, reference is made to a sample of a short paper found in the Appendix. This example has been written in two parallel versions: a "poor" version and a "better" version. Both were created by writers who set out with the same objectives and the same topic. The poor version highlights some of the problematic issues that often diminish the quality of students' writing assignments. The better version applies advice presented in the book and shows how the selected issues were addressed.

Why Write?

You might think that essays and other forms of written work demanded by your professor are torture exercises inflicted on you as a part of some ancient academic initiation ritual. To tell the truth, however, there are some very good reasons to develop expertise in writing:

- It is an *academic and professional responsibility* to write. As an academic or a practising geographer (perhaps a labelled planner, demographer, or geomorphologist) or an environmental scientist, you should make the results of your work known to the public, to government, or to sponsoring agencies. Not only is there a moral obligation to make public the results of scholarly inquiry, but you will probably be required to write—and to write well—as part of any occupation you pursue.

Writing promotes original thought.

- Writing is one of the most *powerful means of communicating* that we have. It is also the most common means for the formal transmission of ideas and arguments.
- Writing is *vital to the development and reshaping of knowledge*.
- Among the most important reasons for writing is that writing is a *generative, thought-provoking process*:

I write because I don't know what I think until I read what I have to say.

—Flannery O'Connor

You write—and find you have something to say.

—Wright Morris

Was it only by dreaming or writing that I could find out what I thought?

—Joan Didion

As the above quotations suggest, writing promotes original thought. Writing also reveals how much you have understood about a particular topic. (For a discussion of ideas on this point, see Game and Metcalfe's [2003] chapter on writing.)

- Through writing, you can *obtain feedback on your own ideas*. Through the circulation of your writing in forms such as professional reports, essays, letters to the editor, and journal or magazine articles, you may spark replies that contribute to your own knowledge as well as to that of others.
- By forcing you to marshal your thoughts and present them coherently to other people, writing is a *central part of the learning process*.
- Writing is a means of *conveying and creating ideas about new worlds.* Writing is part of the process by which we give meaning to the world (or worlds) in which we live. One of the ways that we make sense of our world is through the writing of others (such as journalists). As a result, those who have power over communication have power over thought and, hence, power over reality.
- *Writing can be fun*. Think of writing as an art form or as storytelling. Use your imagination. Paint the world you want with words.

Keys to Good Writing

In order to write well, you need to become informed by collecting evidence and to present your opinion clearly. Given that you may be expected to write a large number of assignments during your program, you may want to spend a little time ensuring that the many hours you spend on those assignments are as productive and rewarding as possible. Here are some keys to that end:

- *Read widely*. Reading the work of others, especially geographers and environmental scientists, is essential. Not only will doing so increase your knowledge of your subject, but it will also allow you to see *how* experts write about it. Consider reading sometimes not for the content of a piece but simply to focus on the writing. To that end, see the Appendix, which gives two alternative versions of a sample paper. Use the better version to detect how problems characterizing the poor version were corrected.
- *Devote enough time to research and writing*. There is no simple formula for calculating the amount of time that you need to devote to writing assignments of any particular length or percentage value within a course. Some writers do their best work under great pressure of time; others work more slowly and may need weeks to write a short piece well. However, regardless of your writing style, doing the research does take time. So, do yourself a favour and devote plenty of time to finding and reading books and articles and to consulting other sources germane to your assignment.

- *Practise writing*. In most art forms, just as in sport, practice improves your ability to perform. Moreover, practice allows you to apply the conventions of effective writing.
- *Be sure that you fully understand the topic and assignment instructions*. Different types of writing assignments have different characteristics and purposes (e.g., describing, presenting an opinion, persuading). Make sure you understand the specific nature of the assignment.
- *Plan*. Plan your work schedule to allow time for writing, and plan the framework of your written work.
- *Write freely at first*. Follow the framework that you created, but for your first draft write freely within that framework. Suspend **editing** until you have a significant first draft of whatever part of your assignment you are working on (Greetham, 2008). Attempting to write a first draft that is polished can lead to writer's block. If you allow yourself to write freely, you will have something to revise or polish, and you can avoid the increasing anxiety of facing a blank page or screen as the submission deadline approaches.
- *Write multiple drafts*. Perhaps you have an image of a good writer sitting alone at a keyboard typing an error-free, comprehensible, and publishable first draft of a manuscript. Sadly, that image is unfounded. Once you have a first draft—something to work with—review and revise it. Repeat the process until you have a polished final draft. Chapter 3 includes a section about reviewing and revising drafts of an essay, suggestions that can be adapted to any writing assignment.
- *Seek and apply feedback*. While writing is an individual process, almost every successful writer produces many drafts and seeks **comments** from peers and other reviewers. Listen to their comments carefully, but remember that, in the end, it is *your* work.
- *Read your "final" copy aloud*. No matter how many times you read over your work in silence, you will almost always be able to improve it further by correcting it after hearing it read aloud. You will hear repetitions you did not know were there. Unclear ideas and sentences will reveal themselves, and pompous, verbose language will become evident.

Write for your audience

In addition to the above points, when you write, remember that you are writing for an audience. Although in reality only one person—your marker—may read your assignment after you submit it, try to envisage a larger audience for your work, which will likely be the situation when you progress to higher levels of academia or to a professional career. Avoid focusing narrowly on how the marker will react as this tends to lead students to write according to what they *think* the marker wants to read rather than following personal paths of inquiry and expression.

It is vital that you understand the ways in which an audience might react to your work. A valuable means of gaining such understanding is by allowing friends, tutors, and others to read and comment on drafts of your work. You might even find it useful to form a group with some friends and agree to read one another's writing critically.

Avoid plagiarism

The recommendation in the bulleted list on page 4 to seek comments on your writing from other people requires a mention of **plagiarism**. We discuss this topic more fully in Chapter 2, but a note of caution belongs here, too. Plagiarism includes submitting work that is not your own without giving credit. In seeking feedback on your work, you should be asking reviewers to comment on aspects that they find particularly in need of attention, such as poor organization, lack of clarity, and confusing arguments. However, comments on the positive aspects of your writing are also beneficial. What you cannot do—without committing plagiarism—is allow someone to rewrite text, write new material, or modify your writing in such a way that it becomes their work and not yours. If they did so, in order to avoid plagiarism you would have to put their name as well as your own on the title page—and how many professors would accept that unless group work was specified in advance?

The point at which plagiarism is considered to have occurred is not always easy to discern. For example, discussing ideas is a central pillar of university education, and professors tend to encourage discussion. So if someone reads your work and suggests that an idea is incorrect or expressed poorly, this may lead to a discussion of the understanding you both have of the idea. Ultimately, if someone helps you by reading and critiquing your work, you have a personal academic responsibility in the end to ensure that the work you submit is your own. As mentioned, further discussion of plagiarism, including an exercise about recognizing both valid and invalid practices, can be found in Chapter 2.

Get help with writing

If you have questions about writing, there are three places to seek help. First, consult writing guides in the library or online. Second, find out if your institution has a writing centre. Third, ask your professor. Although professors and writing-centre staff are often reluctant to do extensive revisions with you, since it can become questionable who is doing the writing, they will offer general guidance and should be able to answer your questions on specific points.

How to Improve the Quality of Your Written Expression

Write fluently and succinctly

Write simply. Short words (that is, words with fewer syllables) and sentences are best. Limit the use of **jargon**. Unless you wish to confuse your readers, there is little room for obscure language in effective written communication. In your work you should be conveying knowledge and information, not showing how many big words you can use. One straightforward means of checking that communication is clear is to ask yourself whether your writing would be understood by someone whose first language is not English. If you know someone in that position who is willing to read your work, give him or her a copy of your paper to look over.

Make every word count. Padding is easily detected, and it makes assessors suspect that you have little of substance to say. Prune unnecessary words and phrases from your work. Remember, the objective is to answer the question or to convey a body of information—not to fill up space to reach the minimum assignment length!

Pay attention to paragraph structure

Effective paragraph structure is important. Although there are exceptions (see Clanchy and Ballard, 1997, for a discussion), Box 1.1 explains that **paragraphs** usually have three parts. In addition, most paragraphs are unified by a *single* purpose or a single theme (Moxley, 1992, p. 74). That is, a paragraph is a cohesive, self-contained expression of one main idea. If your paragraph conveys a number of separate ideas—each requiring its own supporting sentences—rethink its construction. Divide long paragraphs into two or more shorter ones and revise each one according to the structure suggested in Box 1.1.

In the Appendix (p. 271), the first paragraph in the poor version of the sample paper is poorly constructed. It does not have a single purpose. The writer includes various pieces of information about the topic, but the resulting paragraph is too long and lacks unity. Its lack of focus makes it ineffective as an introductory paragraph for the paper. (See more about this topic in Chapter 3.)

Box 1.1 ◉ Parts of a paragraph

- *Topic (or introductory) sentence:* states the main idea (e.g., "The depletion of Brazil's tropical rain forests is proceeding apace"). It can also serve as a transition from a preceding, related paragraph.
- *Supporting sentences:* consist of evidence and examples to support the main idea or to prove the point (e.g., "There is little government action to end land clearance in fragile environments, and private incentives to clear the land remain attractive").
- *Concluding sentence:* lets the reader know that the paragraph is complete. It may summarize the paragraph, echo the topic sentence, or ask a question (e.g., "There seems to be little hope for the sustainability of forests of the Amazon region").

Source: May (2007, p. 54).

Successive paragraphs should be **related** to one another as well as to the overall thrust of the text. Opening sentences can be used to state the paragraph's topic as well as to provide a transition from the previous paragraph. For example, consider transitional phrases such as the following to begin a new paragraph:

- "Another problem with . . ."
- "Elsewhere . . ."
- "On the other hand . . ."
- "Other common . . ."
- "A similar explanation . . ."
- "A significant consequence is . . ."
- "From a different perspective . . ."
- "A number of issues can be identified . . ."

Use correct grammar, punctuation, and spelling

Writing that contains grammatical, punctuation, or spelling errors tends to draw the reader's attention away from the argument you are trying to make or the message you are trying to convey. You can be less concerned about these elements during earlier stages of your writing, but increase the attention you give to them during later stages of revision and proofreading.

One simple way of detecting difficulties with grammar is to have that trusty friend read your work out loud to you. If that person has difficulty and stumbles over sentence constructions, the grammar is likely in need of repair. Another simple means of avoiding problems is to keep sentences short and simple. Not only are long, convoluted sentences often difficult to understand, but they are also grammatical minefields.

Similarly, punctuation errors can cause confusion for your readers, which can lead to overall frustration if they are distracted from being able to focus on evaluating the quality of your argument. See Chapter 6 for tips on correct usage of punctuation; for further assistance, consult a writing guide or seek advice.

Finally, incorrect spelling brings even the best of work into question. In these days when most assignments are composed on a computer, there is little excuse for incorrect spelling. Be sure to use your computer's spell checker before you submit an assignment. Remember also that a word that is spelled correctly but is simply the wrong word for the context will not be identified as erroneous. (See Box 6.3 in Chapter 6 for a list of common errors.) Go back to that friend you have been troubling for assistance and have him or her look over your paper with an eye to any spelling errors. To answer questions about specific matters of style for Canadian writers, look for *The Canadian Style* (Translation Bureau, 1997) in the reference section of your library. For Canadian spelling, refer to a Canadian dictionary.

Spelling errors and sloppy presentation reflect poorly on your work.

Keep your assignment to a reasonable length

A key to good communication is being able to convey a message with economy (consider the communicative power of some short poems, such as *haiku*). Take care not to write more words or submit more pages than have been asked for. Most people marking papers do not want to read more words than necessary.

Do not include irrelevant "filler" material that serves to extend your assignment to the guideline length but does not help make your case or address the question asked. Avoid masking a scarcity of ideas with verbose language. If you find you do not have enough to say without filling space with irrelevant material, you need to do some more research and thinking about the topic before you do more writing.

How to Improve the Presentation of Your Work

Select a brief and informative title

The best **titles** are short, accurate, and attractive to potential readers. When you have finished writing, check that the title is representative of your paper. An example of a functional and informative—though slightly bland—title for a report is "Social Consequences of Homelessness for Men in Calgary,

Alberta (1990–2006)." This title lets the reader know the topic, place, and time period. An example of a poor title on the same subject matter is "Men and Homelessness." Although this title contains two key words about the topic, it is too vague to represent the actual content of the paper.

Compare the titles of the poor and better versions of the sample paper in the Appendix. Remember that the writers of both pieces were working on the same assignment. The title for the poor version is "Ice Jam Floods." For the better version, the writer chose "Flow Direction of Rivers: An Important Control on the Frequency of Ice-Jam Floods." When you read the two versions, consider which title is a more effective, succinct summary of the topic addressed in the paper.

Make sure your work is legible and well presented

> *Poor presentation can prejudice your case by leading the reader to assume sloppiness of thought.*
>
> —Bate and Sharpe, 1990, p. 38

Work that is difficult to read because of poor presentation can infuriate assessors because it is very difficult for them to keep your case, argument, or evidence in mind if they are repeatedly interrupted by distracting elements. Markers now generally expect that post-secondary students will compose their writing on a computer. Use one-and-a-half or double-line spacing because it is easier to read and allows room for the assessor to make corrections or insert comments. Wide margins also provide additional space for comments. However, wide margins or line spacing, or a large font, should not be used to meet a minimum required length if that has been defined by number of pages rather than number of words. Markers will quickly see through this ploy, not only because the wider margins, larger type size, or line spacing is obvious but also because the ideas in your assignment are unlikely to have the expected depth of development.

Students frequently pay little attention to the quality of presentation and, as a result, submit assignments with a sloppy appearance. For that reason, many professors draw up a list of rules for formatting and preparing assignments. For example, you may be required to use a certain font, line spacing, and margin sizes, to follow a specific reference style, to include a separate title page (or not), or to print only on one side of the page. Having a list of rules can help you develop your skills by enforcing good practices. Because it is also easier to follow specific rules than to make all of the formatting decisions yourself, having them allows you to put more effort into improving the substance of your work. To avoid deductions for simple formatting errors, find out if your professor has specific rules. If so, be sure to follow them.

Nicely presented work suggests pride of authorship. You are likely to find that presentation does make a difference—to your own view of your work as well as to the view of the assessor. The first thing your reader will see is the title page, which should include the following:

- The title of your assignment
- Your name and student number
- The name of the professor and the course number
- The date the report was submitted

With environmental awareness increasing, options for using less paper are becoming more common; such options do not detract from the presentation quality of written work. Some of those options are to use line-and-a-half spacing (instead of double), to print double-sided, and to put the title-page information at the top of the first page of text instead of on a separate (and mostly empty) page. If your professor has guidelines that require more paper to be used, consider asking whether the guidelines can be updated to allow for more environmentally friendly practices.

> ***Inquire whether title page information can be shown on the first page of text instead of on a separate page.***

One way to use no paper at all is to submit your assignment in electronic form rather than as a hard copy. Your assessor, however, may prefer a hard copy for the purpose of writing comments instead of typing them into the document using commenting functions. Again, if this method has not been stated as a submission option or requirement, ask your professor if it would be acceptable.

Use headings to show structure

Headings may be useful in several ways (Goldbort, 2006, p. 142). Headings help map the structure of written work; show readers where to find specific information in the work; group information into clearly defined sections; and sometimes **indicate** to the reader what is to follow. Use enough headings to offer a person doing a quick scan of the piece a sense of its structure or intellectual "trajectory." To check whether you have done that, write out the headings you propose to use. Is the list logical or confusing, sparse or detailed? Referring back to the piece itself, revise your list of headings until it gives a clear, succinct overview of your work.

The use of headings is not mandatory for all types of written work. Short pieces are less likely to require headings, for example, and some essay writers view headings as optional. If you decide not to show headings despite the benefits mentioned above, you must be particularly vigilant within the text itself to help the reader identify the structure and follow the trajectory of your work.

If you are writing a longer piece, you may need a hierarchy of headings. Two or three levels should be enough for most purposes, although four levels are **illustrated** in the examples in Box 1.2. Too many headings can cause confusion rather than improve clarity. Placement of a heading can be used to indicate its hierarchical position (e.g., a centred heading has a higher position in the hierarchy than one that is left-justified). Text in upper-case letters is more difficult to read than upper and lower case

Headings are advised in a longer piece to help the reader understand its structure.

Box 1.2 ◉ Examples of hierarchical heading styles

In these examples of two, three, and four levels of hierarchical heading styles, the heading would sit alone on a line. The new paragraph would begin on the line below the heading. Typically, headings are numbered in reports but not in essays.

With numbered headings, shown on the right, you may also use different styles for the different levels. However, because the numbering indicates the level, using both numbers and different styles (as shown in the first two examples) may be viewed as redundant. In the third example, the numerical headings are shown without additional style formatting.

Two Levels of Headings

Bold Upper and Lower Case	**1 Bold Upper and Lower Case**
Italic lower case	*1.1 Italic lower case*

Three Levels of Headings

BOLD UPPER CASE	**1. BOLD UPPER CASE**
Bold Upper and Lower Case	**1.1 Bold Upper and Lower Case**
Italic lower case	*1.1.1 Italic lower case*

Four Levels of Headings

BOLD CAPITALS	1. First Level Heading
Bold Upper and Lower Case	1.1 Second Level Heading
Bold italic lower case	1.1.1 Third Level Heading
Italic lower case	1.1.1.1 Fourth Level Heading

text, so keep those headings short or consider skipping full upper-case headings and start your hierarchy with a style that uses both upper and lower case.

Box 1.2 shows one system of formatting hierarchical headings with numbered and non-numbered options. Essays typically use non-numbered headings; reports, numbered headings. Look at some of your textbooks, and at journal articles and other published materials, and you will see that there are many different systems used. Note the unique formatting methods (e.g., styles of text) that are used to indicate positions in the hierarchy. More important than the specific system used (unless you have been given a style sheet to follow), however, is that you take care to ensure that formatting of section headings is *consistent* throughout the work. Headings that have the same format indicate to the reader a common level in the hierarchical structure of your work. If the formatting is inconsistent, the structure will be confusing or difficult for the reader to follow.

Format source citations and references properly

Here are two pieces of advice on **references** that will help save time during the final stages of assignment preparation. First, when collecting **sources**, keep a full record of **bibliographic details**. If you record only partial information, or none at all, you will have to spend time later looking for the source again just to obtain the missing details. A good way of recording the full record is to keep a copy of the title page of every source you reference (either by saving an electronic copy, printing the page, or photocopying it). If the source is an article, you may be able to save the full document electronically. Chapter 6 describes what constitutes a full record of bibliographic details for different types of references. That chapter also contains information about software for storing and managing bibliographic details and exporting them to a **reference list**.

The second piece of advice we can give regarding references and citations is to be sure to insert your **citations** as you are writing your essay. It is more difficult to come back to a paper and try to insert the correct references afterwards (Hodge, 1994a). By then, you may not remember which source corresponds to which reference or on which source page a particular quotation is located. Searching through references for information you cited just to insert the correct citation in your paper can waste valuable time and increase stress as the submission deadline approaches.

> *Keep a full record of all your sources. It will save you a lot of time and work when you prepare your final reference list.*

Making Sure Your In-Text Citation Style Is Correct and Consistent

The most common—and most easily rectified—problems in written work emerge from incorrect acknowledgement of sources. Full and correct acknowledgement of the sources from which you have derived quotations, ideas, and evidence is a fundamental part of the academic enterprise. Indeed, many markers consider that problems with the relatively simple matter of citations are indicative of more serious shortcomings in the work they are reading. Acknowledging the contributions of others should not be difficult if you follow the instructions on citing sources provided in Chapter 6. Also see the Appendix for examples of how to incorporate citations into a writing assignment, as well as some of the common errors that are made.

> *Citing and referencing sources properly are vital skills to master in academic writing.*

Formatting Your Reference List Correctly

Reference lists in students' written work are frequently formatted incorrectly or inconsistently. An appreciation for the art of creating a reference list that is complete, accurate, and consistent is important. Bibliographic software for creating reference lists can help you prepare a list that is properly formatted. See chapters 2 and 6 for help with citing sources and formatting a reference list. The poor version of the sample paper in the Appendix includes a list of references that contains many errors; a correct list is shown in the better version. Diligence, attention to details, and knowledge of formatting guidelines are the keys to making a good reference list.

Before you submit an assignment, be sure that all in-text citations have a corresponding entry in the reference list at the end. Furthermore, in most cases the reference list should include *only* those references you have actually cited in the paper. This contrasts with a **bibliography**, which is a *complete* list of sources that were consulted during the research and writing process.

Demonstrate a level of individual scholarship

"Scholarship" is one of the most important and perhaps the least tangible of the qualities that make a good assignment. Above all, the assignment should clearly be a product of *your* mind, of *your* logical thought. The marker will react less than favourably to material that is merely a compilation of the work of other writers. An example of such work is an essay that consists of paraphrases and quotations from sources with the writer's contribution being limited to providing linking words and phrases but no original thoughts. In considering matters of scholarship, then, markers are searching for a judicious use of reference material combined with *your individual insights*. Use direct quotations sparingly and be sure to integrate them with the rest of your text. In addition, keep paraphrasing to a minimum. For further discussion of and advice on quoting, paraphrasing, and using sources, see Chapter 2.

Markers want to know what you think and have learned about a specific topic.

You might argue that if you are a novice in the discipline, you must rely heavily on other people's work when writing an assignment such as a research-based essay. Obviously, you might encounter problems when you are asked to write an essay on a subject that, until a few weeks ago, may have been quite foreign to you. But do not be misled into believing that in writing an essay you must produce some earth-shattering exposition on the topic you have been assigned. Instead, your assessor is looking for evidence that you have read on the subject, *interpreted* that reading, and set out evidence based on that interpretation that satisfactorily addresses the assignment topic.

How to Use Supplementary Materials Correctly

Looking through a typical geography or environmental science textbook will reveal that **supplementary** (or illustrative) **materials** (e.g., diagrams, tables, and maps—see Table 1.1 for naming conventions) are commonly used to complement the text and to make points more clearly, effectively, or succinctly than using words alone. Why not follow the models in your textbook and include supplementary materials in your work? People often remember or understand information in illustrative materials better than that in the text. Supplementary materials do not need to come directly from your reading, however. You can create your own supplementary materials to help to convey your ideas where appropriate (see chapters 8 and 9).

Table 1.1 ◉ Supplementary materials and how they are named

Type of Material	Name
Graphs, diagrams, maps	Figure
Tables, charts	Table
Photographs	Figure or Plate

Make effective use of figures and tables

When evaluating your use of supplementary materials, markers will check to see that you have made reference to the materials in your text and that the materials make the points intended. It is valid to emphasize a particularly important element of an illustration when you refer to it in the text, but avoid excessive description or repetition of its details. Markers also look to see whether you might have used more supplementary materials to support the points you are making or to improve the organization or clarity of the information you have presented.

> ***Remember the phrase "A picture is worth a thousand words."***

But take care when incorporating supplementary material into your work. Make sure it is relevant, and locate the material as closely as possible to that part of the text where it is discussed. Unless you have a particularly good reason for doing so, do not put the material in an **appendix** at the end of the document. A reader should be able to evaluate your arguments by assessing the material presented in the main body, including supplementary material that summarizes trends or patterns in larger sets of data. An appendix can be used to provide a more complete listing of data or evidence presented in your work.

Present the supplementary materials correctly

Tables, figures, and plates can contribute substantially to the message conveyed in a piece of work. However, you must take care with their presentation (see Box 1.3).

Titles of supplementary materials should be informative but succinct and accurate. For example, "Chinese-Born Population as Percentage of Total Population in Vancouver, 2012" is a good title. "Vancouver's Chinese Population" is not; it is too vague. If more detailed information is necessary than can be provided in a single title, include a subtitle or a **caption**. Like titles, subtitles should be succinct. Captions, on the other hand, are more comprehensive about the contents or purpose of the figure or table and are often written in full-sentence form. (Note that *title* and *caption* are sometimes used synonymously instead of being distinguished as is done here.)

Detailed information about part of a table may be given in column or row headings, or in footnotes, rather than in a caption. As with other supplementary material, the source from which you derived the material should

Box 1.3 ◉ Using supplementary materials

When using supplementary materials, be sure that they are as follows:

- *Legible.* Material should be large enough to show details clearly.
- *Comprehensible.* Is the illustration easily understood and self-explanatory?
- *Clear.* Photocopied or scanned material can often be blurry.
- *Customized to your work.* Do not submit work padded with marginally relevant photocopied figures or tables taken directly from texts and journals or pasted in from the web. Where appropriate, redraw, rewrite, or modify the material to suit your aims.
- *Identified with sequential numerals.* Give your illustrations a number customized to your work (beginning with Figure 1 or Table 1), and remove all trace of the original source numbering. For example, you should not photocopy Table 17.3 from a book and insert it in a report showing its label as "Table 17.3." If it is the first table in *your* assignment, label it "Table 1."
- *Placed in correct order.* When you include supplementary materials in your work, ensure that they are placed in correct numerical order. For example, Figure 3 should appear before Figure 4. Figure 3 should also be mentioned in the text before Figure 4. If you find that you need to refer to Figure 4 first, switch the labels and change the placement of the figures correspondingly.

Be aware that copying illustrative material directly from a source and inserting it into your work may be a violation of copyright laws. At a minimum, you must clearly identify the source and original number (e.g., "Figure 7.3, Table 2") if you do copy material. Rather than simply copying and risking copyright violation, you should consider recreating the material on your own (but still indicate the number of the material in the source as it is not your original work). In doing so, if you alter the original material to fit your specific purpose, be sure to say, "Modified from [source, figure, or table number]."

be specified with a citation in a line below the figure or table (e.g., "Modified from Jones, 2005, Table 2.4"). Failing to identify the source correctly—or at all—is one of the most common mistakes that students make when using supplementary materials in written work.

Maps and other diagrams should have a complete **legend**, or **key**, that allows readers to understand or decipher the material shown. *Labelling* should also be neat, legible, and relevant to the message being conveyed by the illustration.

For more information on the characteristics and presentation of figures, tables, and maps, see chapters 8 and 9.

Further Reading

Friedman, S.F., and Steinberg, S. (1989). *Writing and thinking in the social sciences.* New York, NY: Prentice-Hall.

- This is a valuable reference on all stages of the writing process.

Game, A., and Metcalfe, A. (2003). *The first year experience: Start, stay and succeed at uni.* Leichhardt, Australia: Federation Press.

- This book has a good chapter on writing that deals gently with the "psychology" of writing as well as its mechanics.

Kneale, P.E. (2003). *Study skills for geography students: A practical guide.* London: Arnold.

- A broad range of information is presented to help students find academic success at university.

Mohan, T., McGregor, H., and Strano, Z. (1992). *Communicating: Theory and practice* (3rd ed.). Sydney, Australia: Harcourt Brace.

- This book provides an overview of the communication process, with some detailed chapters on writing. Emphasis is given to the character of writing and audience responses, and the mechanics of writing particular styles of document (e.g., reports, essays, memos, faxes).

Northey, M., Knight, D.B., and Draper, D. (2012). *Making sense: Geography and environmental sciences* (5th ed.). Toronto, ON: Oxford University Press.

- This is a high-quality guide to research and writing that is intended specifically for geography and environmental science students.

2 Finding, Evaluating, and Using Sources

If I have seen further it is by standing on the shoulders of giants.

—Isaac Newton

Key Topics

- The importance of sources in scholarly writing
- Differentiating academic from non-academic sources
- Finding sources containing useful and relevant information
- Evaluating the credibility of sources
- Incorporating reference material into your writing
- Steering clear of academic dishonesty, including plagiarism
- Preparing a research ethics application or environmental impact statement

Writing in geography and the environmental sciences often requires you to find **sources** and incorporate reference material into your work. This is known as secondary research. (Primary research means that *you* create information—for example, by collecting and analyzing data—rather than acquirie information that *others* created and published.) Material found in secondary sources is important for putting your writing into context and for giving background information about your topic. By using sources writers can, as Isaac Newton wrote, benefit from the body of knowledge generated by previous authors and build upon it. This chapter primarily discusses finding, evaluating, and using secondary sources in your writing. Formatting in-text citations and preparing a list of references cited are dealt with in Chapter 6.

This book is about writing and communication. It is not about the primary research process and it does not intend to address methodologies or analytical techniques. However, you may find yourself in a situation where you plan to collect data or information and are required to prepare a research ethics application or environmental impact statement in advance. These topics are addressed briefly at the end of the chapter.

Why Are Sources Important in Scholarly Writing?

Some proportion of what any person knows is derived informally from personal observation and experience, but most of our formal knowledge has probably been acquired through the educational system. So where does that formal knowledge come from? Over centuries of inquiry, humans have built up a vast quantity of knowledge. Though knowledge can be passed on orally, the most useful way to disseminate it broadly to academic audiences is to record it, most commonly in written form. If we did not tap into the ever-growing body of knowledge (e.g., by reading, discussing, and learning), each person would have to start from the beginning when investigating a new topic. Of course this is impractical and it does not happen because earlier generations have passed on their knowledge to following generations in a continual cycle of learning about a subject, investigating new questions, and recording new findings. By referring to sources, we can present our ideas efficiently, beginning at an advanced starting point. Because the information we acquire comes "second hand" from others as opposed to the information we create "first hand" for ourselves, the term *secondary sources* is used.

Consider a textbook that you are required to buy for a course. In most cases, a professor selects a textbook that represents an up-to-date summary or discussion of the current state of knowledge in the course subject. In addition to his or her insights and opinions, the author of that textbook draws upon sources from the field. These sources will in turn refer to previous knowledge, and so on. The author of the current textbook does not have to trace the development of the entire body of knowledge up to that point, again because it would make for impractical reading. By referring to the book's list of sources, interested readers can search back through the network of **references**; as a result, authors generally need refer only to the more recent advances or state of knowledge. Unlike books of pure fiction, few if any scholarly books stand alone; they are merely the most recent extensions of a body of knowledge.

You may not consider your essay or report to be on par with a textbook, but unless it is a work of fiction it almost certainly has roots in the existing body of knowledge (Harris, 2011). While an essay or review is supposed to present your views and opinions on a subject and while a report should **summarize** what you did for your research or laboratory project, much of what you know about your topic will have been learned from various sources. Therefore, you must draw upon the body of knowledge that is available; put your views, opinions, and findings into the context of what is already

> ***Most academic work has roots in the existing body of knowledge.***

known about the subject; and acknowledge previous ideas and work that are not originally yours.

Take a look at the better version of the sample paper in the Appendix now. The writer did not know much about ice-jam floods before conducting some research on the topic. Focus on how information from sources was used to present the evidence needed to address the thesis statement. As well, **examine** how the writer incorporated those sources into the text using the author–date system of citations as well as a variety of sentence constructions.

In addition to using sources to provide context, Harris (2011) lists five other reasons to seek sources for your writing: (1) to strengthen your argument; (2) to add interest to your work; (3) to find new ideas; (4) to discover controversies or disagreements that may exist in the field; and (5) to help you understand the nature of making reasoned arguments. With these reasons to use sources, the challenge is to find reference materials that are useful and relevant to your topic.

Differentiating Academic from Non-Academic Sources

Before proceeding to **describe** the range of materials you can use as sources of information, let us briefly consider how to differentiate *academic* from *non-academic* sources. Instructors often tell students that they must use academic (or scholarly) sources to support their work and avoid non-academic (or popular) sources, but students are frequently uncertain about what characterizes an academic source. The University of Oregon libraries website (UO Libraries, undated) has a helpful guide for distinguishing academic from non-academic sources. This guide has been adapted for Table 2.1. Be aware that the table has two columns, suggesting mutual exclusivity, but in reality there is a continuum between academic and non-academic sources. An item may not fit exclusively into one category or the other.

Table 2.1 is presented here to prioritize how to identify a source as "academic." We will return to the issue in the next section with a more detailed discussion of how to evaluate the credibility of sources.

Finding Sources with Useful and Relevant Information

So you have been assigned or have chosen a new topic for an essay, research project, or report. Now you need to find some relevant sources. Where do you start: do you turn immediately to Google or another search engine

Table 2.1 ◉ Some criteria for differentiating academic from non-academic sources

Criterion	Academic	Non-Academic
Primary audience	• Intended for scholars, researchers, and practitioners in the field	• Intended for the general public
Author(s)	• Experts in the field • Always named	• Journalists or freelance writers • May or may not be named
Referencing	• Uses citations and presents a list of references, or uses a system of footnotes and/or endnotes for references	• Rarely uses a system for citing and listing references
Editor(s)	• Editorial board of independent scholars, employing a system of peer review	• Editor works for publisher
Publisher(s)	• Often a scholarly or professional organization or an academic press	• Commercial, for-profit enterprise
Writing style	• Assumes a level of knowledge in the field • Usually contains specialized language (disciplinary jargon)	• Easy to read—aimed at the general public • Writing often entertains as it informs
General appearance	• Primarily print with few pictures • Predominantly "black-and-white" • Often contains tables, graphs, and diagrams • Usually few or no advertisements	• Photographs often comprise significant proportion • Glossy and colour • Often sold at newsstands and bookstores

Source: Adapted from University of Oregon Libraries (UO Libraries, undated).

on the Internet? (See Box 2.1 for a discussion of the term *Internet* and its usage.) When seeking evidence to include in your essay or report, do not confine your search to material found on free-access **web pages** accessed from search engines. Extend your research activity beyond the simple act of typing keywords into Google. Although the **web** is a remarkable and increasingly valuable source of information (think, for example, of Google Scholar), using it solely and fundamentally indicates lazy scholarship and neglect of the vast amounts of high-quality material not available online or on free-access web pages:

> The mushroom growth of "databases" on the Internet has given the public at large the conviction that it is not *a* but *the* repository of knowledge—"if you want to know anything about anything, click online." This is a misconception (Barzun and Graff, 2004, p. 39–40).

Box 2.1 Distinguishing between the "Internet" and the "web"

Although the terms *Internet* and *web* are commonly used interchangeably, there is a technical distinction. Strictly speaking, the **Internet** is "vast global network of interconnected computer networks" (Hillstrom, 2005, p. xvi) that includes the **World Wide Web** (or "web") along with other component networks such as *e-mail*, *telnet*, and *ftp*. When a **website** (identified with the URL prefix http://) is displayed on a computer, the user is connected to the Internet but more specifically to the web-network component of the Internet. The use of *Internet* for *web* is widespread among the general public although many people are not aware there is a technical distinction.

As noted in the previous section, you should endeavour to use academic sources for your writing. But how often do the results at the top of a Google search fit the description of academic sources in Table 2.1?

Traditional printed materials (books, reports, scholarly journals, magazines, and newspapers) are increasingly being produced in electronic form, but not all the printed materials in existence are online. There may come a time when we can base all research for a particular exercise on sources that are online; however, this section describes the wider array of resources that are currently available to support scholarly research and will correct the misconception that the Internet contains all the information.

With more information being prepared and distributed in electronic form, the distinction between what is found on the web and what is not is becoming increasingly blurred. For example, books were traditionally printed and distributed only in hard-copy form. Now, however, more and more books are available in electronic format in addition to or in place of the hard copy. So, although you may be able to access the electronic version of a book through your library's website, we will describe this below as getting research information "from a book" instead of "from the web." Other important sources of information that are experiencing a dramatic shift toward electronic distribution are journals, magazines, and newspapers. The distinction between what we consider to be or not to be material found on the web is one that will likely be more and more difficult to define as time passes.

Ensure you have enough sources

Just as there is no answer to the question "How long is a piece of string?" there is no specific number of sources you should consult for any particular kind or length of assignment. For instance, you cannot assume that exactly 12 or 20 references is the right number for a first-year, 2000-word essay or that 30 is the correct number for a third-year, 4000-word essay. The right number of sources is the number required to support the ideas, arguments, and evidence you present in your paper. Do not limit yourself to references drawn too heavily from one source or type of information medium.

Most markers give some weight to the *number and range of references* you have used for your work. This is to ensure that you have established the soundness of your case by considering evidence from a broad range of possible sources. For example, the marker might question an essay examining the consequences of hospital privatization on Canadian rural health-care delivery if the essay were based largely on documents produced by only one political party. A broader range of sources will help you to avoid presenting biased or incomplete views on a topic. Remember, too, that most markers will expect you to have consulted sources other than free-access websites!

Types of sources

There are different types of sources, described next, some of which may be more suitable for certain purposes than others. For example, newspapers and magazines mainly report and discuss current items of interest. In contrast, books tend to give a broader perspective on a subject and might be better sources for background material—such as definitions or a review of the current state of knowledge—that can be used to set up an essay or report.

Whether you visit your library in person or online, it is an important place to find research information. Most of the materials described will be available at the library (or through interlibrary loan). Note that although many new materials in this list are being produced in electronic form, older materials may be available only in the original printed form.

Books

Whether in printed or electronic form, books are an amazing source of information. Too often they are overlooked because of the physical effort required to procure them, especially given the size of the web as an information source. Books may be general overviews of a subject (like a textbook) or more specific discussions of specialized topics. Use your library's index to search for titles or title keywords, subject keywords, and authors' names. Books on similar topics may have similar call (index) numbers and are often shelved close together. Visiting the actual stacks will often reveal several useful books on a topic beyond the initial one you found in the electronic index. If a book seems particularly relevant to your research but is not available at your library, inquire whether you can borrow it from another institution through interlibrary loan.

Scholarly journal articles

Articles in scholarly journals are the best source for the most recent peer-reviewed research information. (See Box 2.2 on the importance of peer review.) Do not expect to be able to search your library's index of print holdings for anything but the *titles* of journals. To search for specific information *within* journals (e.g., keywords in titles, abstracts, and sometimes full text; author's names; and specific date ranges), electronic journal databases are used. Access through the library to full-text electronic articles varies by institution. Electronic access to articles published before the 1990s is often limited, but the JSTOR database specializes in older articles. In addition, Google Scholar can be useful for finding out what articles have been published, though tracking down the full text of articles, rather than just an abstract, may be a challenge. Again, consider placing an interlibrary loan request if you are unable to access an article through your own library.

Government reports

Government departments and agencies often produce and publish public reports of their work. This may include "cutting-edge" research, but it also

Box 2.2 • The importance of formal peer review

One of the mechanisms for trying to ensure that the quality of published work meets an acceptable standard is *peer review*. A piece of work can be immediately viewed with much greater respect if it has been examined by others in the field (that is, by peers of the author) before publication. Peer reviewers, along with editors, help to determine whether a work is publishable and suggest how to improve it. Poor peer reviews may result in work being withdrawn from consideration for publication or being improved significantly before publication. Work that is not peer-reviewed is much more likely to contain errors in fact, unsupported or poorly constructed arguments, or biases. If you can, try to determine if the material you have acquired has been formally peer reviewed.

For most of the sources of information discussed in the text above, there will likely have been at least editorial review, if not true peer review, of the work. Book and journal companies have a self-interest in ensuring the credibility and quality of work they publish. Admittedly, the review process is not foolproof, for erroneous or biased material does sometimes get published. However, you are more likely to encounter non-peer-reviewed work in self-published documents and on websites created by individuals rather than in documents distributed through publishing houses.

includes important but more mundane reporting of statistics and conditions that are valuable for understanding the state of a country, province or state, or municipality at a particular time.

Popular magazines

Articles in professional journals are often viewed as having greater scholarly weight than articles in popular magazines, but magazines such as *National Geographic*, *Maclean's*, and *The Economist* may contain useful and credible summaries or discussions of topics from a broader perspective.

Newspapers

Newspapers contain a range of information, including direct reporting, opinion pieces, and editorials. If you seek current information or views (current at the time of an event, that is) about a topic, do not overlook newspapers. Most print newspapers, such as *The Globe and Mail,* now have concurrent online versions with searchable indexes. Note that many of these online newspapers currently follow a common revenue model that allows visitors free access to a limited number of articles per month; beyond that number, a paid subscription is required.

Theses or dissertations

Students in research-based master's and Ph.D. programs (and many undergraduate honours programs) have to write a thesis or dissertation describing their research and findings. Many universities require a copy of the work to be placed in its library as part of the general collection or in a special archive. Universities usually hold copies of only their own theses and dissertations. However, interlibrary loan services may be able to borrow a copy from the host institution.

Free-access websites accessed from search engines

This category was left to the end of the list intentionally to emphasize that there are many sources of information that you may *not* be able to find on free-access websites (i.e., sites that are not restricted to those with a subscription or membership) in search engine results. Then again, the web is already a vast store of information, and it becomes larger and more important as a research tool every day. The next section discusses the need to evaluate the credibility of your sources, especially websites.

> ***Use sources other than free-access websites.***

All libraries in academic institutions have staff available to help users find relevant information. Most offer extracurricular classes or instructional pamphlets about how to search effectively and how to acquire materials and information (in the library itself and beyond its walls). Use these resources, not only to get help with an immediate project but also to learn how to become an effective, independent researcher.

Evaluating the Credibility of Sources

What would you do if you read the following passage in a book?

> The design of Canada's national flag is a grid of blue dots on a solid yellow background. The United States' flag has three horizontal bars: green at the top, red in the middle, and black at the bottom.

Given your existing knowledge, you would not believe what you just read because you know these to be incorrect descriptions of the Canadian and American flags. You would question the *credibility* of the author of this particular passage. Furthermore, the credibility of *all* of the material in that book should be in doubt. But this example is only partially useful because you were presented with facts that you knew to be incorrect based on your existing knowledge.

What about this statement: "The flag of Estonia has three horizontal bars that are blue, black, and white from top to bottom"? Unless you happen to know what the flag of Estonia looks like, you have three choices for how to respond:

1. *Accept the statement as fact.* If you have no existing knowledge to refute the statement, you could just accept it without any further consideration. But what if the statement were erroneous, as were the previous statements about the Canadian and American flags? Although this is one way that you could respond to published material (and too many people do so without applying **critical reading** skills), *it is strongly advised that you do not unconsciously accept a statement as fact.*
2. *Do some investigative work.* You might consider pursuing various avenues of original or secondary research to find out if a statement is correct or not. Sometimes this is a reasonable choice, depending on the information in question, but usually we do not have the time, skills, or resources to investigate personally the validity of everything we read. Most often the next choice is the best one to make.
3. *Evaluate the credibility of the source.* Someone else has already done the investigative work mentioned in the previous point and published his or her findings. Why do that work again, particularly when the topic under discussion

is not as simple as the design of a flag? If you decide the source is credible and you could defend this decision, then you need to trust that the information is true. Even if you make the choice to do some investigative work beyond the first source, unless you are conducting original research you will need to rely on information found in other sources. Therefore, you still need to evaluate the credibility of those sources.

It is possible that you will get "taken in" occasionally when you rely on a source that you judge to be credible. That is, after evaluating how reputable a source is, you may have good reason to accept the information as being credible and correct, but sometimes it will in fact be incorrect. If, however, you assess the reliability of sources carefully, read critically, and search for confirmation in further sources when things do not seem right, you should be safe most of the time. (For the record, the Estonian flag does have horizontal bars with the colours blue, black, and white [Estonian Tourist Board, undated]).

The example of flag designs is just a simple one used to emphasize an important point. *The reason you are consulting references is probably that you are seeking information that you do not already know* or that you are, at least, unsure about. To summarize, for source material that you cannot easily accept or reject by using your existing knowledge, think back to the three possible responses described above. Should you accept statements as fact without any further consideration? No! Should you repeat lengthy or complex investigative work that someone has already done? Not unless you have lots of spare time or a reason to doubt the accuracy of the work. The best approach is to learn how to question and critically evaluate the credibility of your sources. If you become proficient at doing that, you can use the body of existing knowledge to your advantage and spend your time more efficiently.

Questions to help you evaluate the credibility of sources

How should you evaluate the credibility of your sources? Following are some questions to ask. The answers will not be definitive in all cases, but favourable responses to several questions should lead you to accept a source as credible. Alternatively, a single unfavourable response might prompt you to reject a source. Note the similarity between the answers to these questions and the characteristics of academic or scholarly sources (Table 2.1). Because of the requirements that must be met for a source to be considered academic, those requirements usually mean that a source is also credible.

Make sure all of your sources are credible.

- *Who is the publisher?* Sources issued by a publishing house or a government department have normally gone through an editing process and can be considered more reliable than material that is self-published.
- *Is the author anonymous?* Anonymity should not always lead you immediately to reject a source. Indeed, many valid sources do not list individual authors' names (for example, publications from governments, non-governmental organizations, or businesses) because an author is considered to be an agent or representative of the organization. But if a credible organization is not listed, anonymous sources should generally be used with caution (for example, see Box 2.3 about Wikipedia references).
- *Was the material subjected to peer review?* As mentioned in Box 2.2, peer review automatically elevates the status of a source.
- *Does the author provide supporting evidence and citations?* An argument is more persuasive if it is supported by evidence from the larger body of knowledge (Radford et al., 2002).
- *For a free-access web page, does it pass the credibility test?* Table 2.2 lists a number of questions that can be used to evaluate web pages. Just because a web page appears at the top of a list of search results and seems to make your research an easy task does not eliminate the need to assess its credibility as a source.
- *Is the author objective?* Look for evidence that the author has considered both sides of an argument. Radford et al. (2002) point out that even though everyone has a point of view, objective writing presents a balanced view; therefore, it is important to watch for bias. Consider *anwr.org* (Arctic Power–Arctic National Wildlife Refuge, 2013), a site promoting oil drilling in Alaska's Arctic National Wildlife Refuge. It takes some investigation to find that the site is supported by Arctic Power, a coalition of Alaskan industry groups pressing to open parts of the refuge to oil and gas development. Arctic Power is underwritten by the state of Alaska with funding from the oil industry.
- *Can you detect errors in fact?* Even if you decide to accept the credibility of a source after considering the previous questions, you may still find errors that undermine your confidence in the source.

Table 2.2 offers some suggestions for scrutinizing the credibility of web pages. Just because something is published or is on the web does not mean it is true. To quote Radford et al.,

> Because practically anyone can post and distribute their ideas on the web, you need to develop a new set of critical thinking skills that focus on

Box 2.3 ◉ Wikipedia: Facts or fiction?

One web resource that demands particular scrutiny is Wikipedia. When a term is typed into a search engine, a Wikipedia listing often appears among the first few links in the results. Consequently, students frequently rely on Wikipedia as an easy, packaged source of information, without considering the reliability of the information or the credibility of the author(s). In a chapter called "The Lessons of Wikipedia," Zittrain (2008) discusses this phenomenon and the potential downfalls you face if you use its material indiscriminately. Editing privileges on Wikipedia are available to anyone with Internet access, and the editors are generally anonymous. Can you see the potential for someone to give incorrect or distorted information, either intentionally or unintentionally?

The fact that material can be edited by different people can lead to disputes since not all information consists of simple facts that everyone agrees upon. There are mechanisms in Wikipedia to resolve disputes, but this requires someone to question the material and bring it to the attention of others. The Wikipedia organization expects that over time errors and distortions will be removed, but what if a new Wikipedia page containing errors does not attract other users to put their efforts into verifying and correcting the information?

The Wikipedia organization itself, on its own organization information page (Wikipedia, 2013a), cautions users that information may be unreliable. In particular, newer information may be less reliable than older material since incorrect material is more likely to be changed by subsequent editors. However, users should always question whether Wikipedia information has been subjected to adequate measures to ensure accuracy. A rule of thumb for using this resource safely is this: *Use Wikipedia as an initial source to acquire some basic knowledge about a topic, but seek other sources to confirm the accuracy of the information.*

Many Wikipedia pages do follow good practices by providing linked citations to original source materials. Follow these links to investigate the credibility of the original sources. You may find that the Wikipedia authors failed to interpret or present the material accurately or to represent different sides of an issue fairly. Having found and evaluated the original source yourself, if you find it credible you will be able to cite that source rather than the secondary Wikipedia page. You may also find that the Wikipedia information is accurate, but you will then have your own independent corroboration to support that conclusion.

Wikipedia editors are anonymous: do not present a Wikipedia page as a primary reference.

In summary, although much of the information on Wikipedia is undeniably correct, a user does not know which information is reliable. And because the editors are generally anonymous, Wikipedia pages violate accepted guidelines for what constitutes a credible source. Therefore, avoid giving references to Wikipedia pages as primary sources in your essay—find other sources in whose credibility you have greater confidence.

Table 2.2 ◉ Evaluating web pages

Question	How Do You Answer It?
Did you find the page or site through a sponsored link?	Some search engines will direct you to sponsored links first. These are usually identified as such.
From what domain does the page come?	Look at the URL (Uniform Resource Locator). Is the domain name, for example, • commercial (.com or .co)? • educational (.edu)? • governmental (.gc or. gov)? • non-profit (.org)? • miscellaneous (.net)? New domains, such as .tv, are in the process of being introduced.
Who or what organization wrote the page and why? Is the organization reputable?	Good places to start are the banner at the top of the page and any statement of copyright, which is usually located at the bottom of the page. Or look for information under links associated with the page and characteristically entitled "About us," "Who we are," or "Background." Details about the authoring organization are sometimes located in the URL between the http://statement and the first / (forward slash), or immediately after a www statement (e.g., www.statcan.gc.ca refers to Statistics Canada and www.usgs.gov refers to the U.S. Geological Survey). Try gradually truncating the URL to find out about the authoring organization. That is, starting from the end of the URL, delete, one by one, each segment ending with a / (forward slash) until you reach the **home page**. This may reveal new web pages that provide insights into the origins of the page you planned to use. It may also be useful to enter the authoring institution's name into a search engine. This may reveal other information that points to funding sources and underlying agendas.
Is this someone's personal web page or part of a **blog**?	Look for a personal name in the URL. Blogs are often identified as such through their title (e.g., Wired Campus Blog); their URL, which may contain the word *blog*; or in web-page text introducing the blog.
When was the page created or last updated?	Look at the bottom of the web page for a "created on" or "last updated on" statement.
Are the content and layout of high quality?	Check to see if the page looks well-produced and if the text is free of typographical errors and spelling mistakes. Where possible, confirm the plausibility and accuracy of data or other information presented by comparing it with other good sources.

Source: Adapted from Beck (2009) and UC Berkeley (2012). The UC Berkeley web page, in particular, offers useful and detailed advice on assessing web pages.

What Does It Imply?
Information distributed with commercial intent is not always balanced.
Consider the appropriateness of the domain to the material. Is this kind of agency a fitting one for the material being presented?
Web pages are written with intent or purpose—and it is not always the best intent!
Information presented on a personal page or blog is not necessarily a bad thing. However, you will need to find out whether the author is a credible source. Use a good search engine to look up the author. If you cannot determine the author's credentials or professional role, think seriously about whether you should use information from the website.
Old pages may contain outdated information. In almost every case, undated statistical or factual information should not be used.
Poorly laid out web pages do not necessarily contain inaccurate information, but they should cause you to question the author's meticulousness in information gathering and presentation efforts.

the evaluation and quality of information, rather than be influenced and manipulated by slick graphics and flashy moving java scripts (2002, p. 38).

Table 2.2 encourages you to be diligent in searching for the author to whom you should attribute information on websites. Some digging with the suggested tactics should allow you to make a connection in most cases. If not, Turabian (2013) suggests being particularly cautious about using web sources with no identifiable author, publisher, or sponsor (including Wikipedia pages—see Box 2.3): "This makes them the equivalent of any other anonymous source, unlikely to be reliable enough to use without serious qualification" (p. 137). You may find it valuable to revisit Box 2.2, which discusses the concept of peer review.

Box. 2.4 discusses a specific example: evaluating the credibility of web pages on the topic of Canada's controversial National Energy Program. Box 2.5 addresses the issue of open-access publishing.

Box 2.4 ◉ Evaluating the credibility of web pages: An example

For this example, imagine that you have been assigned an essay that requires you to investigate the National Energy Program (NEP) in Canada. The NEP was introduced by the Liberal government in the early 1980s. Many in western Canada, particularly in Alberta, viewed it as an unjustified transfer of wealth from the resource-rich west to the politically dominant east. Commentators (e.g., Perrella, 2009) have attributed to the NEP certain political shifts in Canada that included a feeling of alienation in western Canada and a mistrust of the central Canadian power base.

A web search for "National Energy Program" elicits a variety of websites that mention or discuss the NEP. We will compare four websites that provide summaries of the NEP. The first result is a link to Wikipedia (2013b); the second, to The Canadian Encyclopedia (Bregha, 2012); the third, to the Alberta Online Encyclopedia (Vicente, undated); and the fourth, to FreeAlberta.com (undated).

The use of Wikipedia pages is discussed in Box 2.3. A better choice than Wikipedia can be made to serve as a source about the NEP.

The Canadian Encyclopedia (TCE) is a product of contributions from "more than 5000 scholars and specialists from every discipline . . . from every region of Canada" (Canadian Encyclopedia, 2011, unpaginated). Authors of articles are named, but biographical and contact details are not shown (for privacy reasons). The TCE is a project of Historica Canada and includes a full-time editorial staff. Overall, the TCE can be judged to be a credible source.

Like the TCE, the Alberta Online Encyclopedia (AOE) is the creation of a professional organization, the Heritage Community Foundation. The author of the

Incorporating Reference Material into Your Writing

After collecting and reading your sources, you need to incorporate relevant reference material into your writing as evidence to support your views. Before examining techniques for doing so, we will discuss the principle of acknowledging sources of that material. Details about the specific mechanics of formatting citations according to different referencing systems are provided in Chapter 6.

Acknowledge your sources

Ideas, facts, and **quotations** *must* be attributed to the sources from which they were derived (unless the ideas are your own, of course). Failure to acknowledge sources is a common and potentially dangerous error in student essays (see Burkill and Abbey, 2004). Serious omissions may in fact constitute **plagiarism** (see the discussion of *common knowledge* and plagiarism later in

article about the NEP is named. However, the AOE summary clearly emphasizes the Alberta perspective and the negative reaction that the NEP received from Albertans. This is not to suggest that the AOE is unable to provide an accurate summary of the NEP from an Albertan perspective; nevertheless, careful assessment is required to ensure that the view it presents is not biased..

A review of the NEP on the FreeAlberta.com website has a clear bias. For example, one heading is "How the NEP Destroyed the Albertan Economy and Albertan Jobs." The article's author is not named on the page, and there is limited information available on the website about who makes up the organization. In addition to presenting facts about the NEP that are similar to those available from other sources, this web page contains rhetoric that is meant to inspire support for secession of Alberta from Canada. While the website author(s) would likely argue that their presentation and *interpretation* of facts are correct, the organization does not hide its contempt for the NEP and the federal government. Therefore, the credibility of this website as a primary source for an unbiased review of the NEP is poor. (On the other hand, if your topic was about political discontent in Alberta, this website could be a valuable example!)

A writer is not restricted to choosing one source as a reference on a subject. In this example, both the TCE and AOE websites provide accurate information, but the TCE gives a more neutral summary of the topic. Unlike Wikipedia and FreeAlberta.com, whose use as primary sources has been discouraged, the AOE web page is not necessarily unacceptable as a reference. From the point of view of the independence and credibility of the author and organization, however, the TCE should be considered the first choice among these four for an unbiased review of the NEP.

Box 2.5 Open-access publishing

An important issue that has developed in recent years is the rise of the *open access* movement. Open access represents a major shift in the publishing model of academic journals. In the traditional model, end users pay for the costs of publishing journals, through either institutional or individual subscriptions. Objections have been made about rising subscription costs, leading many subscriptions to be cancelled. In the full open-access model, journals can be accessed for free because the publishing costs are transferred to others. For example, authors are commonly required to pay fees to have articles published. Hybrid open-access models also exist, such as paid access for a limited initial period and delayed release to open access.

The open access movement has risen in conjunction with growth of the Internet. Previously, academic journals were used in print form, and access was limited to institutional libraries or to individuals willing to pay the subscription fees. With the advent of the Internet and the ability to disseminate materials digitally, anyone with Internet access could be a potential reader, except that access remained subscription based. Open access is promoted as a means of eliminating barriers to academic journals for any interested reader.

Many funding organizations require publication of funded research in open access journals so that the results are freely available. Compared to the traditional hard-copy publishing model, publishing on the Internet is easier. Correspondingly, growth in the number of open-access journals has been rapid, which has caused concerns to be raised. Proponents of the open-access movement emphasize that most open-access journals are still peer reviewed, but editorial oversight and the quality of articles in open-access journals have been questioned. Some have asked whether the author-pay system leads to the publication of articles that would not have been published under the traditional model. Nick (2012) wrote a summary and discussion of the main issues relating to the open-access movement. Debate about the benefits and potential drawbacks of open-access publishing is strong, and many other points of view can be found on the web.

Peer review remains a critical aspect differentiating the quality and reliability of information in academic journals, whether they are open-access or not. Open-access *journals*, however, are different from open-access *repositories*, which are digital storehouses into which people can deposit a broad range of peer-reviewed and non-peer-reviewed materials. Caution should be exercised with a source found in an open-access repository; you are advised to check whether such a source went through a peer-review process.

this chapter). **Acknowledgment** should be made by an acceptable reference system. To prevent uncertainty about what is meant by "acceptable," your professor may specify that you use a certain system.

All this business of acknowledging sources may seem to you to be a painful waste of time. But, as with most things, there are reasons for it (see Box 2.6).

Box 2.6 ◉ Reasons for acknowledging sources

People writing in an academic environment acknowledge the work of others for three main reasons (Harris, 2011):

- To attribute credit (and sometimes blame) for the acknowledged author's contribution to knowledge.
- To avoid committing plagiarism. Giving credit appropriately takes care of the plagiarism problem, coincidentally.
- To allow interested readers to pursue a line of inquiry should they be stimulated by something that has been cited. For example, in an essay on the economic geography of winemaking in the Niagara region, someone might have written this: "Baker (2012, p. 16) observed that there is a great deal of money to be made by private investors by investing in the winemaking business." *Aha*, you think, *I can make some quick dollars here. All I need to do is find Baker's book and read about how this might be done*. Thanks to the appropriate acknowledgement of Baker's idea in the essay you have read, you can rush to track down Baker's words of wisdom on money-making. Lo and behold, you become an instant millionaire!

Paraphrase and quote your source material

There are two principal ways of incorporating the material collected from sources into your work: **paraphrasing** and **quoting**. Either way, acknowledgement of the source material must be given with a citation, for the reasons discussed in Box 2.6. Quite often, people new to essay writing incorrectly place all of the citations of references contained in a paragraph at the end of that paragraph. There they place a string of names, years, and page numbers. This is incorrect because the reader has no way of establishing which specific ideas, concepts, and facts are being attributed to whom. Each reference citation should be placed as close as possible to the idea or illustration to which it is connected.

Paraphrasing and quoting are techniques for incorporating source material into your work, but acknowledgment must be given to the original author.

Paraphrasing is the act of representing someone's work (for example, words, ideas, figures, tables, and so on) accurately *without* quoting them directly. At its best, paraphrasing *integrates* the essence of an author's work into your own writing, thus strengthening your point or position while acknowledging the author. At its worst, a writer takes a sentence from a source, replaces some key words or changes the structure of the sentence, and adds a citation in the "new" sentence. Has the source material really been integrated into the writing? In essence, the original material has been copied, although with the changes technically the new sentence is not identical to the original. In summary, good paraphrasing requires more internal processing than just changing the order of someone else's words and adding a citation. Integrate the material into your work using your own wording.

There are two ways of structuring sentences containing paraphrased material: *passive* and *active*.[1] For the following examples, suppose that you have been reading an article written by Michelle Johnson, published in 2008, in which Johnson presents the results of her study of the changes observed at Shady Glacier between 1995 and 2005. In the examples that follow, in-text citations are shown in the author–date system.

- *Passive paraphrasing*. The citation to the reference source is included in parentheses, most often at the end of the sentence containing the paraphrased material. For example:

  > The margin of Shady Glacier retreated 120 metres between 1995 and 2005 (Johnson, 2008).

  For this sentence to qualify as valid paraphrasing, this information must appear in the article but the sentence must not appear as written here. The exact wording must be yours, with the citation (Johnson, 2008) indicating that the information mentioned is not your own. In this example, the short sentence might serve as your summary of a main finding that the author discusses and analyzes to much greater extent. If the sentence *does* appear in Johnson's (2008) article exactly as written above, then it would have to be presented as a quotation (discussed below).
- *Active paraphrasing*. The author(s) of the paraphrased text can be used as the subject of the sentence, followed by a verb such as *stated*, *proved*, or *emphasized*. The past tense is usually used because the work was completed in the past. Using an active structure, the same paraphrase as above could be written as:

  > Johnson (2008) showed that the margin of Shady Glacier retreated 120 metres between 1995 and 2005.

  or

  > As Johnson (2008) showed, the margin of Shady Glacier retreated 120 metres between 1995 and 2005.

[1] This meaning of *active* and *passive* should not be confused with the active and passive voice of verbs, which we discuss in Chapter 11.

The active structure is used less frequently, especially by inexperienced writers. Try using both structures to liven up your writing. Readers can get bored reading text where only the passive structure is used and where citations always appear at the ends of sentences.

Use active and passive sentence structures when paraphrasing.

If you are *quoting* someone *directly*, follow these four golden rules:

1. *Reproduce the text exactly*. Spelling, capitalization, and paragraphing must mirror those in the original source. If a word is misspelled or if there is an error of fact, reproduce the original text but put the word **sic** in square brackets immediately after the error ([these] are square brackets). This lets your reader know that the mistake was in the original source and that you have not misquoted or made a typing error. Your main text and the inserted quotation should be grammatically consistent. This sometimes requires that you remove or add words. If you find it necessary to omit unnecessary words from the quotation, use three periods (. . .), known as **ellipsis points**, to show that you have deleted words from the original text. If you need to add words for your sentence to be grammatically correct, put those you have added within square brackets. Make sure, however, that you do not change the original meaning of the text through your omissions or additions.
2. When making a direct quotation of fewer than about 30 words, you should *incorporate the quotation* into your own text, indicating the beginning and end of the quote with **quotation marks**. The in-text citation or numerical reference (see Chapter 6) is usually placed after the closing quotation marks, as in this example:

   > He described Hispaniola and Tortuga as densely populated and "completely cultivated like the countryside around Cordoba" (Colon, 2006, p. 165).

   If, on the other hand, you are making a direct quotation of more than about 30 words, you should not use quotation marks but, rather, *set apart the whole quotation from the main text and use single-spacing* as shown in the example here:

   > One historian makes the point clear:
   >
   > > Although it included a wide range of human existence, New York was best known in its extremes, as a city capable of shedding the most brilliant light and casting the deepest

> shadows. Perhaps no place in the world asked such extremes of love and hate, often in the same person (Spann, 2001, p. 426).

A blank line immediately precedes and follows the quotation.

3. *Use quotations sparingly*. Because someone else wrote the words in a quotation, you cannot expect to get much credit for it beyond making an appropriate selection for the context and integrating it well into your writing. Quotations should be used much less frequently than paraphrasing; with the latter the writing, at least, is (or should be) predominantly in your own words. Use a quotation only when it is important for the reader to see the exact original wording or when the quotation contains a major statement that you must document. Do not use quotations as a shortcut to replace your own ideas or interpretations.
4. *Integrate the quotation into your text*. **Justify** its inclusion. Let the reader know what it means for your work.

Common knowledge does not need to be cited

Some writers take the advice on acknowledging sources too literally. For example, you might argue that apart from knowledge acquired from direct learned experiences, everything you write must be cited because it has been acquired from sources over time. There is a useful provision that allows people to write without citing everything: **common knowledge** does not need to be cited. Common knowledge "represents the kind of general information found in many sources and remembered by many people" (Harris, 2011, p. 19) and includes easily observable information, commonly reported facts, and common sayings (Harris, 2011). This implies that common knowledge is the same for everyone, but for an individual writer it can be difficult to decide if a specific example qualifies as common knowledge or if a citation is needed. One person's opinion of what is easily observable or commonly reported may differ from others' opinions. Also, what might not be common knowledge at one time may become so later as the information is disseminated.

Common knowledge does not need to be cited—the challenge is to assess what counts as common knowledge and what does not.

Harris's statements may be viewed as broad guiding principles that still require careful consideration in individual cases. The following guidelines may help in situations where the definition and descriptions above are not sufficiently helpful or in situations that are not black and white:

- If you write something that you clearly did not know before you read it in a source, cite the source.
- If you write something without consulting a source, it is probably common knowledge.

Complications arise in the second case when you write after reading about the topic at hand, and your knowledge is then a mixture of recently acquired information and older information. In that case the question to ask might be this one: "Even though I can write this without consulting my source directly, is it reasonable to say that I knew this before I had read the source?" If not, cite the source. Finally, if you present common knowledge in the form of a quotation from a source, you must use quotation marks and give a page reference.

As mentioned above, writers can overdo the number of citations they include. Text can become clumsy and difficult to read with excessive, unnecessary citations. Furthermore, the reader (who may be a marker) may begin to question whether you have demonstrated a sufficient level of individual scholarship (described in Chapter 1) if nearly everything you wrote is attributed to others. When you read the work of other people, take note of where and how often the authors give citations.

Deciding what information to cite and not to cite is ultimately the author's responsibility. Try to strike a balance: give credit where it is due but avoid citing when you can present a credible and defensible argument that it is common knowledge. But if you are uncertain, it is better to cite the source than not.

Steering Clear of Academic Dishonesty, Including Plagiarism

Academic dishonesty can take various forms; it may include the following:

- Failing to acknowledge sources of information or materials used, or plagiarism (see below for an expanded discussion of plagiarism and how to avoid committing it);
- Fabricating or falsifying data or the results of laboratory, field, or other work;
- Accepting assistance with a piece of assessed individual work from another person, except in accordance with approved study and assessment provisions;
- Giving assistance, including providing work to be copied, to a person undertaking a piece of assessed individual work except in accordance with approved study and assessment provisions;

- Submitting the same work for more than one course unless the professors have said that this procedure is acceptable for the specific piece of work in question;
- Purchasing an essay (e.g., from another student or from an online "paper mill": Cvetkovic, 2004) and submitting it as your own work.

Correct referencing is an important academic and professional courtesy. Failing to acknowledge sources of ideas, phrases, text, data, diagrams, and other materials is *plagiarism* and is widely regarded as a serious contravention of intellectual etiquette. In fact, committing plagiarism may lead, eventually, to penalties ranging from minor (e.g., a mark of zero on the assignment) to severe (e.g., expulsion from a course or from the institution).

Plagiarism (from the Latin word for *kidnapper* [Mills, 1994, p. 263]) is a form of academic dishonesty involving the use of someone else's words or ideas as if they were your own. Plagiarism may result from deliberate and deceptive misuse of another person's work, or from ignorance or inexperience about the correct way to acknowledge other work. It can take different forms, including the following:

- Presenting extracts from books, articles, theses, other published or unpublished works (such as working papers, seminar or conference papers, internal reports, computer software, lecture notes or tapes, numerical calculations and data) or the work of other students without clearly indicating the origin of those extracts by means of quotation marks and citations;
- Paraphrasing sentences or whole paragraphs without due acknowledgement in the form of citations to the original work.

Note that cutting-and-pasting material from a digital document or electronic source is *no different* from actually rewriting or retyping someone else's words and submitting them as your own. Whether you actually write or type the words or not, it is the *act* of presenting someone else's work without acknowledgement that constitutes plagiarism. In an electronic age, the ease of cutting-and-pasting can make the temptation to take such shortcuts greater than ever before (Bowman, 2004a).

Cutting-and-pasting instead of rewriting or retyping material still constitutes plagiarism if the source is not acknowledged.

Burkill and Abbey (2004) suggest some useful ways to help avoid plagiarism in your work. First, make it a habit to record all bibliographic details on notes from your reading and on photocopied articles and chapters. Second, and similarly, keep full bibliographic details, not just the website URL, of all material you obtain from the Internet. Third, provide a full list of references, corresponding to the in-text citations, at the end of your written work.

As mentioned, academic institutions regard academic dishonesty as a very serious matter and will usually impose penalties according to policies that are laid out in official documents (e.g., the academic calendar). New technologies to detect plagiarism are being refined. Ignorance of what constitutes plagiarism is generally not regarded as a valid excuse. Do take the time to become familiar with procedures for citing your sources. As a matter of instructional practice, professors may give their students a list of referencing style guidelines to follow. If you are not given guidelines, however, the onus is still on you to acknowledge your sources accurately. Even though your assignment may not explicitly tell you to do so ("But it didn't say I had to . . ."—professors hear all sorts of excuses!), you must still cite and list any sources you use.

Keep in mind that plagiarism is not always as black-and-white as copying material from a published source without giving acknowledgement. Chapter 1 suggested that you ask classmates for criticism of your assignment. While there are clearly benefits to having someone other than yourself read your work with a fresh eye, allowing a person to do work for you (e.g., rewriting or writing new material) and then presenting it as your own work is a violation of the principles of academic honesty.

Box 2.7 presents some hypothetical cases that you might encounter as a student. You must determine whether the actions constitute plagiarism.

Box 2.7 ◉ Thinking about plagiarism

We tend to like questions that are easy to answer, but plagiarism is often a complex issue that appears in shades of grey rather than as a stark choice between black and white (Bowman, 2004a, Eisner and Vicinus, 2008). Considering examples of various situations is beneficial for learning about good practices in writing and avoiding plagiarism. Think about the following six situations and discuss them with your friends and classmates. No answers are given; the intention is to encourage you to acquire a strong understanding of when plagiarism occurs, to examine each situation carefully from different perspectives, and to understand how difficult it can often be to recognize the boundary between plagiaristic and non-plagiaristic acts.

1. Karen has found some relevant information in an article from the journal *Geographical Research*. She copies out a couple of sentences word for word, includes quotation marks, and gives a citation. Is this the work of a plagiarist?

Continued

2. Bill is searching the Internet for inspiration for his essay on "flooding in the Saguenay region of Quebec." He stumbles across a website that contains essays written by other students and finds an essay that is very similar to the topic he's been assigned. He decides to use some of the main points from the essay on the website in his essay without mentioning where they came from, but he does not copy any text directly. Do you think this is plagiarism?
3. You are working in the environmental sciences lab. It's all going well, but then some of the equipment breaks. It will now take ages to obtain the last batch of results. Your friends did the same experiment last week and you ask them to give you their results. Would using their results constitute plagiarism?
4. Four students have worked together on a fieldwork project, and they have been asked to submit individual accounts of the work. Each student submits work with whole sections that are almost identical. Is this plagiarism?
5. Joe waits until the final week when a major assignment is due before he gets down to work. He rather manically grabs some books on planning and constructs a quite substantial amount of his project by piecing together sections from different sources. In-text citations to the original sources are given and he uses his own words to link the sections together. Do you think this is plagiarism?
6. A geography student discovers that two of her classmates who are international students are copying parts of books verbatim, and without attribution, for a semester assignment. When she confronts them, she discovers they do not believe they are doing anything wrong. They claim that in their home country it is understood their responsibility is to gather information from sources and organize it to form their assignment, and that copying to accomplish this objective is acceptable. What do you think?

Source: Adapted from Burkill and Abbey (2004, p. 441) and Preston (2001, p. 102).

If you have any questions about plagiarism that are not answered in this book, take a look at Burkill and Abbey's (2004) article on avoiding plagiarism or at fuller discussions in books such as Bowman (2004b) or Eisner and Vicinus (2008). Alternatively, have a conversation about the issue with your professor. Learn about the topic and be prepared to engage in an **argument** about a case (with yourself or with others) rather than expecting a simple yes or no answer to a question about plagiarism.

Preparing a Research Ethics Application or Environmental Impact Statement

This chapter has focused on the acquisition and use of information contained in *secondary* sources because the information comes from the work of others. In contrast, a *primary* source of information is one that you create yourself, such as the results of a research project. Chapter 4 focuses on the task of writing a report about primary research you have conducted. Here we address an important topic related to conducting research for the purpose of producing a **primary** source of information: a requirement to prepare a research ethics application or environmental impact statement for certain studies.

If you are conducting independent research that involves human or animal subjects, be aware that you will most likely have to write a research ethics application *before starting any data collection*. Most academic institutions have policies that require all researchers—undergraduates, graduate students, and professors—to obtain approval from an internal research ethics board if human or animal subjects are involved. Canada's three main research funding councils—the Natural Sciences and Engineering Research Council (NSERC), the Social Sciences and Humanities Research Council (SSHRC), and the Medical Research Council (MRC)—have developed research ethics guidelines. Failure to obtain research ethics approval may have serious consequences for your institution. As a student you are a representative of the institution so you have a responsibility to meet the requirements, no matter how benign you consider your research or how minor a figure at the institution you perceive yourself to be.

If proposed research activity involves human or animal subjects, research ethics approval must be obtained before starting data collection.

You will be asked to prove that your research methods are designed to ensure that your subjects will be treated according to accepted ethical guidelines during *and after* the data collection period. Be sure to allot time in your research plan to complete the application form and to wait until approval is given before proceeding. Although you may consider your study to be straightforward and clearly within ethical guidelines, do not assume that approval will be granted automatically; and do not begin your data collection prematurely. Your application may be rejected or you may be required to modify your proposed methods. If you are unsure whether approval is required for your study, consult with your professor or the research ethics board. Montello and Sutton (2006) have a chapter dedicated to ethics in research.

Research ethics guidelines have broad applicability. Sometimes even research that may seem to be beyond the need for ethics approval will need approval. The key issue is whether human or animal subjects are involved. Consider the practice of using interviews of human subjects as a means of collecting information. If the questions you intend to ask of the subjects are about personal matters, it is probably easy to see that the ethics of what questions you ask and how you will protect the subjects' privacy must be considered. However, even if you want to interview a person about matters that are *not* of a personal nature, a human subject is involved and, therefore, ethics approval must still be obtained. For example, you may want to ask an expert about the history of pollution and changes in regulatory policy in your region. The expert's opinion would be a professional one, not strictly personal. As a result, you might think there would no ethical issues to be addressed, but you should not assume this to be the case.

The solution to ensuring you have met research ethics guidelines is to obtain information and ask questions. Acquire a copy of your institution's policies and guidelines. Read them carefully and assess whether your research requires prior approval. If you are uncertain, seek guidance. At worst, you can prepare an application and let the committee decide whether or not seeking approval was necessary.

When it comes to collecting information by interviewing subjects and using it for your research, high ethical standards must be maintained. Recording an interview is advisable rather than taking notes, but be sure the subject is aware that you are doing so: ask for permission. Having a recording will enable you to listen to the full interview again later when you can transcribe the entire conversation or make notes from it. A recording can also protect you if the subject claims that he or she said something different from what you report. Playing back the recording would eliminate doubt about what was said.

By agreeing to be interviewed, your subject is placing his or her trust in you, so be sure to report or use material in a manner and context that reflects his or her responses or opinions honestly. A recording can protect you from allegations of misrepresenting responses, but only if you have reported accurately. Use direct quotations. Do not modify responses to suit your own agenda, even if doing so would help your research objectives. Where possible, minimize the subjective aspects of an interview by transferring information into an objective matrix for further analysis of combined data from your full sample group.

Take care to report a subject's interview responses accurately, even if it complicates your research objectives.

Environmental impact statements are similar in purpose to research ethics applications. If you plan to conduct research that could potentially harm or have a lasting impact on the environment, anticipate the need to prepare a statement in advance. You will be required to show how you plan to avoid harm or lasting impact, and what measures you would take in the event of an accident. Demonstrating insufficient awareness or inadequate planning would likely result in rejection of your research project. That, however, is better than seeing an unapproved project go awry and having to face the ensuing remediation and consequences—possibly legal ones.

Further Reading

Barzun, J., and Graff, H.F. (2004). *The modern researcher* (6th ed.). Belmont, CA: Thomson Wadsworth.

- This book provides an in-depth discourse on research skills emphasizing that, in the "modern" age, not everything can be found on the Internet.

Beck, S.E. (2009). *Evaluation criteria.* http://lib.nmsu.edu/instruction/evalcrit.html (accessed 20 November 2009).

- Here you will find a very helpful description of prewriting and writing strategies and enjoyable reading.

Bowman, V. (ed.). (2004b). *The plagiarism plague.* New York: Neal-Schuman.

- In addition to describing the background to the issue of plagiarism, the author suggests remedies for the problem and recommends resources to consult for guidance.

Eisner, C., and Vicinus, M. (eds.). (2008). *Originality, imitation and plagiarism.* Ann Arbor, MI: University of Michigan Press.

- Not a reference book, this edited volume is about teaching writing in a digital age and is intended for professors. Some chapters may be of interest to anyone wanting a deeper understanding of originality, imitation, and plagiarism in student writing.

Harris, R.A. (2011). *Using sources effectively: Strengthening your writing and avoiding plagiarism* (3rd ed.). Los Angeles: Pyrczak.

- This book provides a close examination of the use of sources. It contains many examples and exercises for improving your skill in using sources and testing your understanding of using sources.

Montello, D.R., and Sutton, P.C. (2006). *An Introduction to scientific research methods in geography.* Thousand Oaks, CA: Sage.

- Chapter 14 discusses research ethics in detail.

Radford, M.L., Barnes, S.B., and Barr, L.R. (2002). *Web research: Selecting, evaluating, and citing.* Boston: Allyn and Bacon.

- This is a short but useful book about web research. Chapter 3 is about content evaluation of websites, and Chapter 5 discusses copyright in connection with the use of material from the web.

University of California at Berkeley. (2012). *Evaluating web pages: Techniques to apply and questions to ask.* www.lib.berkeley.edu/TeachingLib/Guides/Internet/Evaluate.html (accessed 18 November 2013).

- This is a very helpful and practical resource for those seeking detailed guidance on assessing the quality of web pages.

3 Writing an Essay

To my mind, the best essays are deeply personal (that doesn't necessarily mean autobiographical) and deeply engaged with issues and ideas. And the best essays show that the name of the genre is also a verb, so they demonstrate a mind in process—reflecting, trying-out, essaying.

—Robert Atwan

Key Topics

- Quality of argument
- Quality of evidence
- A self-assessment form for your essay

Most students will encounter some form of **essay** assignment regularly during their academic careers. Essays are favoured by professors because they require a combination of elements: research, investigation, reflection, and organization of your ideas on the topic; and clear expression of those ideas in written form. Not all essays are identical in their objectives, however: see Box 3.1 for a description of different types of essays you may be required to write. *Essay* is used here as a general term to cover anything from a short opinion piece to a full-length term paper.

This chapter is devoted to examining what your essay markers might be looking for when they are **assessing** your work. In addition, the chapter discusses the steps involved in writing a thesis statement and creating an essay structure or framework. Good planning before you actually begin writing will help you to clearly express your thoughts and ideas. Much of the information and advice in the following pages is framed around the essay assessment form included at the end of the chapter.

Students often say, "I don't know what my professor is looking for with this assignment," as if the task were to duplicate as closely as possible some mythical perfect essay that their professor already has in mind. However, professors are not looking for "correct answers" in an essay. There is no predetermined

Box 3.1 ◉ Types of essays in geography and the environmental sciences

An essay, according to one definition, is simply "a written composition on a specific subject" (May, 2007, p. 72). Essays in geography and the environmental sciences are often expected to be more than *descriptive* or *expository* reviews that recount all the facts the author has discovered about some phenomenon or issue. A more applicable definition of an *academic* essay is a "composition giving the writer's views on a specific subject or expressing his or her opinions on a specific subject" (May, 2007, p. 72). This chapter assumes that your essay topic requires you to present an informed opinion or analysis of a topic, that is, an opinion or analysis based on evidence you have **synthesized** from reference sources.

Some essay topics require that you do more than just present your informed opinion to the reader. In an *argumentative* or *persuasive* essay (Buckley, 2009), for example, the author makes an **argument** to persuade the reader of a particular position on a subject. Make sure that you read the assigned essay topic carefully, identify key terms, and understand exactly what you are being asked to do. The process of breaking down the essay topic during the planning stage and reviewing the topic during the revision stage is examined later in this chapter.

There are other, closely related, types of academic writing that are not called essays. Chapter 4 discusses *reports,* in which the writer summarizes the findings of some original research. *Summaries* and *reviews* are discussed in Chapter 5; these compositions provide the reader with a synopsis of a written work and, in the case of a review, also give the writer's evaluation, or opinion, of the work. A longer review, furthermore, may be called a review essay or a *synthetic critical* essay, particularly if a number of sources are summarized and evaluated in an integrated manner. In this case the writer's opinion focuses on the sources (such as the quality of discourse, logic, or argument) rather than on the subject itself.

"line" for you to follow. They are concerned with your perspective on a question and how well you make your case. Presenting incorrect facts or faulty reasoning will result in poor evaluations, but whether the marker agrees or disagrees with your judgment is not essential to your mark. Disagreement does not lead to bad marks; bad essays do (Lovell and Moore, 1992).

As Box 3.2 shows, the answer to the question "What is your marker looking for?" is really quite simple.

Box 3.2 ◉ What is a marker looking for in your essay?

An essay marker wants . . .

- to be told clearly what you *think*
- and what you have *learned*
- about a *specific* topic.

The following guidelines and advice, which are written to match the criteria outlined in the assessment form at the end of this chapter, ought to help you satisfy the broad objective noted in Box 3.2. It is worth thinking seriously about these guidelines.

Quality of Argument

Ensure that the essay addresses the question fully

If any one issue in particular can be identified as crucial to a good essay, it is this one. Failure to address the question assigned or chosen is often a straightforward indicator that the student does not understand the course material. It may also be seen as a sign of carelessness in reading the question or of a lack of interest and diligence.

- Look closely at the wording of your essay topic. What does **describe** mean? How about **analyze** or **compare** and **contrast**? What do other key words in the assigned topic actually mean? The meaning of a topic's key words can be important to the way you approach your essay. For example, an essay in which you are asked to "critically discuss David Suzuki's role in raising public awareness of environmental issues in Canada" requires more than an uncritical description of what he has done. The wording asks you to **evaluate** the significance of his actions. Similarly, an essay that asks you to "discuss the impact of tourism on Canada's national parks" requires that you focus on the impact of tourism and not that you dwell on a long historical summary of facts and figures about tourism in Canada's national parks.
- Discuss the topic with other people in your course. See how your classmates have **interpreted** it. Listen critically to their views, but be prepared to stand up for your own and change them only if you are convinced you are wrong.
- Whenever necessary, clarify the meaning of an assigned topic with your professor. Do this *after* you have given the topic full thought, discussed it with friends or classmates, and established your own interpretation, but before you begin writing your paper. Meeting with a student who says he doesn't understand the topic when it is apparent that he has not really thought about it first is frustrating for professors. In these situations, the student is not really saying, "I don't understand the topic"; he is saying, "I want you to do the thinking for me."
- When you have finished writing, check that you have covered all the material required by the nature of the topic. The technique of listing

and rearranging paragraphs, which is outlined in the next section of this chapter, may be helpful.

Be sure that your essay is developed logically

> Nothing is more frustrating than to be lost in someone else's intellectual muddle. A paper that fails to define its purpose, that drifts from one topic to the next, that "does not seem to go anywhere," is certain to frustrate the reader. If that reader happens to be your [professor], he or she is likely to strike back with notations scrawled in the margin criticizing the paper as "poorly organized," "incoherent," "lacking clear focus," "discursive," "muddled," or the like. Most [professors] have developed a formidable arsenal of terms that express their frustration at having to wade through papers that . . . are poorly conceived or disorganized (Friedman and Steinberg, 1989, p. 53).

In assessing an essay, markers will usually look for a coherent framework of thought underpinning your work. They are trying to uncover the conceptual skeleton upon which you have hung your ideas to see if it is orderly and logical. Throughout the essay, readers need to be reminded of the connections between your discussion and the framework. Make clear the relationship between the point being made and any argument you are advancing. For advice on developing a structure to frame your essay before you begin creating sentences and paragraphs—an important part of good writing—see "Creating a Structure for your Essay," later in this chapter.

Write a thesis statement

As well as developing a structure or framework for your essay, you should write a thesis statement. This may be done prior to developing the structure—it might be your first action after collecting and organizing your source materials—or concurrently with the development of the structure. A *thesis statement* is a concise summary of what you intend to argue, prove, or achieve in your essay. This message should be set out clearly near the beginning of your essay, typically as part of the introductory section. An effective thesis statement will give your reader the impression that your essay will be focused and organized, not aimless and incoherent. A clear thesis statement shows that you understand the topic or assignment objectives.

> ***A thesis statement indicates what you intend to argue, prove, or achieve in your essay.***

A thesis statement should reflect your ideas or opinions, whether these are personal ideas or opinions or academic ideas and opinions synthesized from your research. It should go beyond simply repeating a sentence found in a handout that broadly describes the assignment objectives. Write the thesis statement after you have given your topic sufficient thought and have done your research.

In many cases a thesis statement can be written concisely and effectively in one sentence. However, you are not constrained to use only one sentence—remember, the term is thesis *statement*: a statement of your objective or objectives for the essay. If you need to write a couple of sentences or even a short paragraph to express your idea clearly, do so.

A thesis statement does not have to be as explicit as "This essay will prove that . . ." or "The purpose of this essay is to argue for. . . ." Here are two examples of effective thesis statements that are presented more subtly:

> The problem of deposition from acid precipitation in North America has been reduced in severity since 1970.

> The problem of deposition from acid precipitation in North America has been reduced in severity since 1970 primarily because of the United States' *Clean Air Act*.

At first glance these statements may seem the same but there is an important difference. In the first essay, the writer's objective is to provide evidence to show that there has been a reduction in the severity of acid precipitation deposition in North America since 1970. The second statement is more targeted. The writer needs to present some contextual evidence that deposition from acid precipitation has been reduced, but the main thrust of the essay will be to prove that the reduction can be attributed to the introduction of the *Clean Air Act*.

To illustrate further what distinguishes a good thesis statement, consider the thesis statements written for the sample paper in the Appendix. In the poor version, the following sentence at the end of the (long) first paragraph was intended to be the thesis statement: "This paper is about ice-jam floods and why they are common in the Northern Hemisphere." Contrast it with the thesis statement in the better version: "Examining the nature and causes of ice-jam floods reveals why northward-flowing, higher-latitude rivers in the Northern Hemisphere are particularly susceptible to these events." The latter statement provides the reader with a more specific idea of what the paper will do (examine the nature and causes of ice-jam floods) and what it sets out to achieve (to explain why northward-flowing, higher-latitude rivers in the Northern Hemisphere are particularly susceptible to ice-jam floods).

Throughout the essay your guiding principle should be to develop the argument or provide the evidence necessary to support your thesis statement. In the concluding section, you should remind the reader of your thesis statement, but with a new perspective. Guide the reader to reflect on the information you presented to support the thesis statement. Think of a tarpaulin laid on the ground on a windy day. You can stake down one side, but the other side will be free to flap in the wind unless you stake it down as well. By including a thesis statement in the introduction and then revisiting it in the conclusion, you stake down both sides of your argument, ensuring a smooth, coherent essay.

Create a structure for your essay

Some methods of developing an essay structure are outlined in Box 3.3.For most topics an essay framework can be formed before writing begins (see "Essay outline" and "Sketch diagrams"). In other situations, the essay may take shape as it is being written (see "Free writing").

Plan your essay.

Composing an outline with introduction, body, and conclusion

This section elaborates on writing an essay based on a formal **outline** with headings. In almost all cases, good academic writing will have an **introduction**, a main **body**, and a **conclusion**. To visualize this structure, you might imagine an essay as taking the form of an hourglass. The introduction provides a broad outline, setting the topic in its context. A reader provided with a sense of direction early in the paper should find your work easy to follow. The central discussion tapers inward to focus on the detail of the specific issues you are exploring. The conclusion sets your findings back into the context from which the subject is derived (by revisiting the thesis statement) and may point to directions for future inquiry. In a longer essay, the introduction and the conclusion may each need many paragraphs rather than the single paragraphs that might suffice in a shorter essay.

Make sure your essay structure is clear and coherent.

The guidelines in Box 3.4 are not suggested as a strict recipe for essay writing, but they should be of some assistance in constructing a good paper. The sample paper in the Appendix, although short, illustrates how they can be applied. The better version contains a clear introduction (one paragraph), body (three paragraphs), and conclusion (one paragraph). The poor version does not have a distinct introductory paragraph as it mixes an overall introduction to the paper's topic with details more suitable for the body.

Box 3.3 ◉ Forming an essay structure

Forming a structure for your essay is often the best way to make the writing flow more easily. Instead of staring at a blank page, seeing the structure you intend to follow will help you to write a first draft, which can then be revised and improved upon.

Essay outline

Try to prepare an **essay plan** or **outline** before you begin writing. That is, work out a series of broad headings that will form the framework upon which your essay will be constructed. Then, add increasingly detailed headings and ideas for individual paragraphs. What will be the main ideas you might cover and in what order? What examples, data, and quotations might be critical to supporting your thesis statement? What conclusions might you reach? As you proceed, you may find it necessary to make changes to the essay's overall structure.

Sketch diagrams

A sketch diagram of the subject matter is also a good means of working out your essay's structure. Write down key words associated with the material you will discuss, and sketch out the ways those points are connected to one another. Rearrange the diagram until you have formulated a structure that can then be transferred into the essay outline format with headings described above.

Free writing

If you encounter "writer's block" or are writing on a topic that does not lend itself initially to an essay plan, try brainstorming: without hesitation, write anything related to the topic until you have some paragraphs on the screen or page in front of you. This is **free writing**, which is different from *writing freely,* in which, as we saw on p. 4, you write within the framework of your plan but without interrupting the flow by stopping to revise. Free writing is unconstrained, less disciplined, and more exploratory (Academic Skills Centre, 1995; May, 2007). It is a way of unlocking ideas, making connections between elements of the material you have read about, and potentially removing obstacles blocking your writing progress. What you have written by using free writing may be of quite poor quality structurally, but if you previously found it difficult to create an essay plan, the ideas you uncover may help to define your plan.

Alternatively, you may create something that you can work into a first draft. After free-writing some material, remove the "junk" and organize what is left into a coherent package. Look for unnecessary repetition and for opportunities to move sentences and paragraphs around. To be effective, this writing style requires a good knowledge of the material to be discussed. It is *not* an easy option for people who do not have a clue about their essay topic.

Box 3.4 ◉ Guide to the introduction, body, and conclusion of an essay

Follow these guidelines for your *introduction*:

A good essay gets off to a good start.

- **State** your aims or purpose clearly. What problem or issue are you discussing?
- Include a *thesis statement* (described in the previous section). What case will you argue?
- Make your **conceptual framework** clear. This framework gives the readers a basis for understanding how the ideas that follow fit together in a coherent manner.
- Set your study in context. What is the significance of the topic?
- Delineate the scope of your discussion (that is, give the reader some idea of the spatial, temporal, and intellectual boundaries of your presentation).
- Give the reader some idea of the plan of your discussion (Buckley, 2009). This will prepare readers for the intellectual journey they are about to undertake.
- Be brief. In most essays an introduction that is about 10 per cent of the total length of the essay is adequate.
- Capture the reader's attention from the outset. Is there some unexpected or surprising angle to the essay? Or catch his or her attention with relevant and interesting quotations, amazing fact, and amusing anecdotes. Make your introduction clear and lively because first impressions are crucial.

The best introductions are those that get to the point quickly and that capture the reader's attention (May, 2007). Like a good travel guide, an effective introduction allows readers to **distinguish** and understand the main points of your essay as they read through them (Greetham, 2008).

Follow these guidelines in the *body:*

Convince your reader with logic, example, and well-organized structure.

- Make your case. Answer the academic equivalent of the central question in a crime mystery: "Who dunnit?"
- Provide the reader with reasons and **evidence** to support your views. If it is an argumentative essay, make sure your argument is persuasive. Imagine your professor is sitting on your shoulder (an unpleasant thought!) saying "**prove** that" or "I don't believe you." Use your writing to disarm him or her.
- Present your material logically, precisely, and in an orderly fashion.
- Accompany your key points with carefully chosen, colourful, and suitable examples and analogies.
- In high school, students are often taught to write essays according to the following formula: introduction, body with three distinct points, and conclusion. For that reason some students think that all essays must always contain three points: no more and no fewer. While the strict three-point

Continued

model may be suitable, the valuable lesson is to get used to creating the *general* structure—introduction, body with sufficient evidence to address the essay question, conclusion.

Follow these guidelines in your *conclusion*:

- Tie the conclusion neatly together with the introduction. When you have finished writing your essay, read just the introduction and the conclusion. Do they make sense together? Finally, ask yourself, "Have I answered the question?" and "Have I supported my thesis statement?" The answers may be evident only if you have the opportunity to reread your essay some days after you have finished writing it. For this reason, if for no other, it is a good idea to plan to finish the second-last draft of your essay some days before the due date. This will give you the opportunity to review your work in a relaxed state of mind.

Make sure the conclusion matches up with the introduction.

- Do more than simply restate the essay question. Use the conclusion to highlight the main points you made in the essay (May, 2007) and to tie your ideas together so that the reader is not left dangling (Northey et al., 2012).
- The conclusion ought to be the best possible summary of the evidence supporting your thesis statement that you have discussed in the main body of the paper (Friedman and Steinberg, 1989). It must match the strengths and balance of material you have presented throughout the essay (Greetham, 2008).
- If relevant, discuss the broad implications of the work (Buckley, 2009).
- Do not introduce new ideas or any points not mentioned previously.

Avoid clichéd, phony, or overly dramatic conclusions (Northey et al., 2012), such as "The tremendous amount of soil erosion in the valley dramatically highlights the awful plight of the poor farmers who for generations to come will suffer dreadfully from the loss of the very basis for their livelihood." Instead, try to leave your reader with something interesting to think about (Najar and Riley, 2004), such as your informed perspective on recommended changes or future trends (Johnson, 1997).

Review and revise your essay drafts

If you think you can produce a finished product in one go—the first draft—it is unlikely that you are producing high-quality writing. For most writers the first draft may take the longest time to write, but the writing process does not end there. With a first draft in hand you can move on to reviewing it and making the revisions necessary to raise your essay to a level that is suitable for submission. Be careful to set aside the time required to conduct effective reviewing and revising; these steps cannot be done at the last minute. Indeed, the first draft must be completed well in advance of the submission deadline to allow sufficient time to prepare subsequent drafts. There is no rule for how many drafts should be written, but each draft should be an overall improvement on the previous one. As you progress from first draft to final copy, there will be fewer major changes to make and more minor polishing can be done.

As mentioned at the beginning of the Appendix, the poor version of the sample paper might be an example of what a writer comes up with on the first draft. In that light it is not too bad—as long as it not left there. It provides a start. Most writers find it easier to revise something already written than to create the first draft. By reviewing that first draft and using it as a base for subsequent revisions, the writer of the poor version of the sample paper could easily develop it into the better version.

Checking the structure

When you have written the first draft of your essay, assess the structure. It may align well with the outline you created before you started writing. Or, while writing, you may have deviated from your initial plan. If so, the structure in the draft might be acceptable or you may need to consider rearranging sections. You can do this quite easily:

- Read through the headings. Are they in a logical order? Do they address the assigned topic in a coherent fashion? Did you omit any section that you had planned to include? If the headings differ in order from your outline, determine whether the initial plan was better and if sections should be rearranged. If necessary, rearrange the sections until they do make sense, and add new headings that might be necessary to cover the topic fully.
- If additional headings are required, you will also have to write some new sections of your paper. Of course, you may also find that in contrast to your initial ideas, some sections are unnecessary and can be removed.
- Make your revisions and then go through this process of assigning and arranging headings again until you are satisfied that the essay follows a logical progression.

Remember the advice from the beginning of the chapter. Your marker wants you to **explain** clearly what you think and what you have learned about a specific topic.

Ensuring the material is relevant

The material you present in your essay should be clearly and explicitly linked to the topic being discussed. To help clarify whether material is relevant or

Don't waffle!

not, try the following exercise. Go through the first draft of your essay, read each paragraph, and ask yourself two questions:

1. Does *all* of the information in this paragraph help answer the question posed?
2. *How* does this information help answer the question?

On the basis of your answers, revise. This should help you to eliminate irrelevant material.

Ensuring the topic is dealt with in depth

Have you simply slapped on a quick coat of paint, or does your essay reflect preparation, undercoating, and a good final coat? Have you presented arguments that are persuasive? Have you explored all of the relevant issues emerging from the topic? This does not mean that you should employ the shotgun technique of essay-writing, indiscriminately putting as much information as you can into the pages of your essay.

Instead, be diligent and thoughtful in going about your research. Take notes or make photocopies. Read. Read. Read. There is not really any simple way of working out whether you have dealt with a topic in sufficient depth. Perhaps all that can be said is that broad reading and discussions with your professor will provide some guidance.

Reading and revising from different perspectives

During one revising session, focus on big-picture questions. Does the essay make sense overall? Is there good organization and flow? Is your argument logical and supported by sufficient, appropriate, and balanced evidence? In later revising sessions, when you are getting closer to submitting the essay, narrow your focus to more detailed aspects, such as sentence construction, spelling, and punctuation.

Another, very effective, means of checking the fluency of your writing is to put a draft of your work away for several days and then read it afresh. Odd constructions and poor expression that were not evident before will leap off the page. This exercise is often even more revealing if you read the essay aloud. Or you could make a deal with some friends: you will "evaluate" their papers if they will "evaluate" yours. We tend to see other people's mistakes more easily than our own because we are more familiar with our own style and material.

Reading your essay aloud can reveal clumsy expressions, poor punctuation, and repetition of words or sounds.

Quality of Evidence

Ensure your essay is well supported by evidence and examples

You need relevant examples, statistics, and quotations from books, articles, and interviews, as well as other forms of evidence to support your case and substantiate your claims. See Chapter 2 for guidance on finding and evaluating sources. Any evidence or examples that are drawn from sources other than personal experience must be cited and referenced properly. (Chapter 6 provides information on citing and referencing techniques.)

In addition, most readers seek examples that will bring to life or emphasize the importance of the points you are trying to make. You can draw information from good research (e.g., by reading widely). Careful use of examples is also an indicator of diligence in research and of the ability to link **concepts** or theory with "reality."

Used judiciously, personal experience can be good evidence for an essay.

Personal experience and observations may be incorporated into written work as evidence. For example, a police officer with several years of experience may be able to offer penetrating insights into an assignment on the geographies of crime and justice. Women and men who have spent time caring for children in new suburban areas may have valuable comments to make on providing social services to such areas. Referring to your own experience is certainly valid when relevant, but be sure to clarify in the text that you are doing so, and give the reader some indication of the nature and extent of your experience: e.g., "In my nine years as a police officer in Toronto . . ." However, use personal observations sparingly and not as a substitute for actual research-based evidence. Where possible, support your personal anecdotes with related evidence from other sources that readers will be able to consult.

Avoid making unsupported **generalizations**, such as "Crime is increasing daily" or "Air pollution is the major cause of respiratory illness." These statements are not necessarily incorrect, but they do need to be supported. Unsupported generalizations are a sign of laziness or sloppy scholarship and will usually draw critical remarks from essay markers. Support your claims by using empirical evidence or citing recognized sources, e.g., "Environment Canada has recently released a report entitled *Canada's 2007 Greenhouse Gas Inventory—A Summary of Trends* (Environment Canada, 2008) that **demonstrates** that total greenhouse gas emissions in 2007 were about 26 per cent higher than 1990 levels"; or "A paper (Kunzli et al., 2000) in the medical journal *The Lancet* suggests that 40 000 deaths annually in Austria, France, and Switzerland can be attributed to air pollution."

Presenting your evidence and examples accurately

When you use examples, take care to ensure that they

- are relevant,
- are as up-to-date as possible,
- are drawn from reputable sources and identified with citations, and
- include no errors of fact.

Keep a tight rein on your examples. Use only those details you need to make your case.

A Self-Assessment Form for Your Essay

Box 3.5 is an example of a structured form, or *rubric*, that an assessor might use to evaluate your essay and assign an overall mark. If the professor makes an evaluation form available before you submit your essay, you can use it during your revisions to guide improvements and address specific points in your writing.

Box 3.5 ◉ Essay assessment form

Student Name: Grade: Assessed by:

This form provides an itemized rating scale of various aspects of the written assignment. Sections left blank are not relevant to the attached assignment. Some aspects are more important than others, so there is no mathematical formula connecting the scatter of ticks with the final grade for the assignment. Ticks in either of the two boxes left of centre mean that the statement is true to a greater (outer left) or lesser (inner left) extent. The same principle applies to the right-hand boxes. If you have any questions about the individual scales, comments, final grade, or other aspects of this assignment, please see the assessor indicated above.

Quality of discourse

Essay addresses the question					Essay fails to address the question
Essay is developed logically					Writing rambles and lacks logical continuity
Writing is well structured through introduction, body, and conclusion					Writing is poorly structured, lacking introduction, cohesive paragraphing, and/or conclusion
Material is relevant to topic					Not all material is relevant to topic
Topic is dealt with in depth					Superficial treatment of topic

Quality of opinion, argument, and evidence		
Opinions are clear, arguments are convincing	☐☐☐☐	Opinions not expressed clearly, arguments are not convincing
Opinions or arguments are well supported by evidence and examples	☐☐☐☐	Inadequate supporting evidence or examples
Accurate presentation of evidence and examples	☐☐☐☐	Evidence is incomplete or questionable
Effective use of figures and tables	☐☐☐☐	Figures and tables rarely used or not used when needed
Illustrations presented effectively and cited correctly	☐☐☐☐	Illustrations presented poorly or cited incorrectly

Written expression and presentation		
Fluent and succinct piece of writing	☐☐☐☐	Writing is clumsy, verbose, or repetitive
Sentences are grammatically correct	☐☐☐☐	Sentences contain grammatical errors
Correct punctuation is used	☐☐☐☐	Punctuation is poor
Correct spelling is used	☐☐☐☐	Spelling is poor
Work is legible and well presented	☐☐☐☐	Work is untidy and difficult to read
Length is appropriate for assignment guidelines	☐☐☐☐	Essay is too short or too long

Sources and references		
Adequate number of references used	☐☐☐☐	Inadequate number of references used
Acknowledgement of sources is adequate	☐☐☐☐	Acknowledgement of sources is inadequate
Correct and consistent in-text citation style is used	☐☐☐☐	In-text referencing style is incorrect or inconsistent
Reference list is presented correctly	☐☐☐☐	Reference list contains errors or inconsistencies

Continued

Demonstrated level of individual scholarship

High | | | | | Low

Assessor's comments

Further Reading

Academic Skills Centre. (1995). *Thinking it through: A practical guide to academic essay writing* (revised 2nd ed.). Peterborough, ON: Trent University.

- This book is a helpful guide to the progression of steps involved in writing an academic essay. Chapter 6, "Prewriting," discusses various activities that can precede the actual writing of a first draft, e.g., generating ideas, creating an outline, and free writing.

Buckley, J. (2009). *Fit to print: The Canadian student's guide to essay writing*. Toronto, ON: Nelson Education.

- This book is a detailed source for learning more about the sequence of steps involved in essay writing.

Dixon, T. (2004). *How to get a first: The essential guide to academic success*. London, UK: Routledge.

- Be sure to read the sometimes entertaining chapter on planning an essay, which stresses the importance of both thinking and planning before writing.

Greetham, B. (2008). *How to write better essays* (2nd ed.). Basingstoke, UK: Palgrave Macmillan.

- Here you will find useful material on writing effective paragraphs.

Lester, J.D. (1998). *Writing research papers* (9th ed.). New York, NY: Longman.

- This book includes extensive coverage of the essay-writing process, from finding a topic to writing a proposal, doing library research, writing notecards, and eventually writing the paper.

May, C.A. (2007). *Spotlight on critical skills in essay writing*. Toronto, ON: Pearson Education Canada.

- This is another informative general guide to essay writing that contains useful examples and exercises.

Northey, M., Knight, D.B., and Draper, D. (2012). *Making sense: Geography and environmental sciences* (5th ed.). Toronto, ON: Oxford University Press.

- A high-quality guide to research and writing, this book is intended specifically for geography and environmental science students.

Schwegeler, R.A., and Shamoon, L.K. (1982). The aims and process of the research paper. *College English, 44*(8), 817–824.

- The authors distinguish between, and discuss the implications of, students' views of essays (as an opportunity to show how much "good" information you have collected and presented according to academic conventions) and academics' views of essays (as an opportunity to analyze, interpret, and express an argument).

Shields, M. (2010). *Essay writing: A student's guide*. Los Angeles, CA: Sage.

- A comprehensive and well-received recent guide, this book deals with a broad range of issues involved in essay writing, including planning, gathering information, and citing references.

4 Writing a Research Report

Research is the process of going up alleys to see if they are blind.

—Marston Bates

The great tragedy of science—the slaying of a beautiful hypothesis by an ugly fact.

—T.H. Huxley

Key Topics

- Why write a research report?
- What are readers looking for in a research report?
- Writing a research report
- Writing a laboratory research report
- Writing a research proposal

Your professor has assigned you a research project and, as if that were not difficult enough, you are also required to report on your work—in writing. This chapter, about writing research reports (including theses) and laboratory research reports, is intended to help you perform that task. Indeed, by offering some guidance on how to write a particular kind of report, the following pages may also help you to undertake the research. Reports differ from essays primarily in that the former are based on original research and are mostly objective in nature, whereas the latter are expected to present evidence from secondary research to support the author's informed opinions on the topic. In some situations you may be required to prepare a research proposal. Although it would be written *before* undertaking the research, the chapter concludes with a section on the advanced topic of writing a research proposal.

Why Write a Research Report?

There are at least three good reasons for learning to write research reports. These range from the practical to the principled.

First, there is a *vocational reason*. Academic and professional writing often involves the communication of research findings (Goldbort, 2006). As Montgomery (2003, pp. 138–139) notes, "because of the tendency to outsource analysis and research these days, particularly in industry, the number and diversity of technical reports have gone through a burst of expansion." Thus, people working in geography and the environmental sciences can expect to undertake research and to write associated reports in the course of their employment. Getting and keeping work in these areas often requires the effective conducting and communication of research. And that communication is usually to an audience that expects answers to a certain set of questions irrespective of the character of the project. Consequently, you must be familiar with the ways in which research results are customarily conveyed from one person to another (that is, the conventions of research communication).

The second reason for learning to write research reports is because research reports and articles are a *fundamental and increasingly important building block of knowledge*. Each report is the final product of a process of inquiry that might be a simple lab experiment or in-depth research, such as an honours thesis project or an independent study. Reports are common devices for communicating findings in the professional sphere (e.g., by industry consultants and government researchers). Moreover, through the communication of research findings we contribute to the development of "practically adequate" knowledge of how the world works—that is, knowledge that will not necessarily be absolutely and forever correct. Instead, such knowledge works and makes sense for practical purposes here and now. Some future event or discovery, however, may date that knowledge or replace it.

The third reason to learn to write research reports is that there is a *moral responsibility to present our research honestly and accurately*. Representing the world to other people in ways that we understand it is to play an enormously powerful role. To a degree, people entrust researchers with creating knowledge. Given that trust, our actions in conducting research and reporting on the results must be beyond reproach. In partial acknowledgement of that trust, we must provide peers, colleagues, and interested observers with accurate accounts of our research activity and results. Therein lies a vital role of research reports and a most important reason for writing them well.

What Are Readers Looking for in a Report?

Research and laboratory reports usually answer five classic investigative questions (see Box 4.1, after Eisenberg, 1992, p. 276).

Box 4.1 Five investigative questions that characterize research reports

- What did you do?
- Why did you do it?
- How did you do it?
- What did you find out?
- What do the findings mean?

The person reading or marking your report seeks clear and accurate answers to these questions. Because reports are sometimes long and complex, the reader will also appreciate some help in navigating the document. Make the report clear and easy to follow through easily understood language, a well-written introduction, suitable headings, and, if appropriate, a table of contents.

Compare the five questions of research reports, as illustrated in Box 4.1, with the purpose of an essay. An essay, as discussed in Chapter 3, generally should present your informed *subjective* opinion about a topic after assessing material collected from secondary research. In contrast, reports generally present the findings of primary research or investigation conducted by the author. The questions outlining the main contents of a report ask for more dispassionate, *objective* responses and do not demand a statement of your personal opinion. Some parts of a report may, however, include "professional" opinions (rather than purely personal opinions). Indeed, recommendations to fit a particular situation that are drawn from analysis of research results often require some professional opinion, particularly if there are several valid choices that could be recommended. However, subjective opinions are usually limited in extent and should be restricted to the discussion, conclusion, and recommendations.

Some forms of the report, especially lab reports, will answer the five investigative questions through a highly structured progression (e.g., introduction, methods, results, discussion, conclusion) and be written in a way that would allow another researcher to repeat the study independently. For example, an environmental scientist reviewing levels of contaminants in the St. Lawrence River or a demographer conducting a statistical study on population patterns in Japan is likely to conduct the study as impartially as possible and to record his or her research procedures in sufficient detail to allow someone else to reproduce the study. For such forms of inquiry, which involve **quantitative methods**, repetition is an important means of verifying results. (This notion of reproducibility, or **replication**, is discussed fully in Montello and Sutton, 2006.)

In other forms of research, such as those involving **qualitative methods** (i.e., interviews, participant observation, or textual analysis), results are confirmed in different ways. As a consequence, the research report may be written differently. It will usually answer the five questions listed in Box 4.1, but less emphasis will be given to the business of ensuring replicability. It is more important that qualitative research reports be written in a way that allows other people to confirm the reliability of your sources and to check your work against other related sources about the same or similar topics. Consider, for example, the way a murder trial is conducted. The murder itself cannot be repeated to allow us to work out who the murderer was (replication). Instead, the lawyers and the police assemble evidence to reconstruct the crime as fairly and accurately as possible. This process is known as **corroboration**. Your report should be a fair and reasonable representation of events. (For fuller discussion of these and related matters, see Mansvelt and Berg, 2005, and DeLyser and Pawson, 2005.)

The great diversity of research topics found in geography and environmental science means that you are likely to be asked to write research reports of different types throughout and after your degree program. Yet almost all of these will usually still answer the five basic investigative questions, even though style and emphases may make the reports quite different in presentation. For that reason, this chapter provides an introduction to report writing *in general*, with some specific references to writing laboratory reports.

It is helpful to have an example of a high-quality report that satisfies the requirements of your professor or whoever else "commissioned" the report (Montgomery, 2003, p. 141). You can use this as a general model of the kind of style you should be adopting or as a precise blueprint, depending on your audience's expectations. As with essays, professors who have specific notions about structure and formatting may give you their own detailed guidelines. If you follow the guidelines carefully, you will be less uncertain about preparing the report, and you will be able to devote your attention to improving its content and presentation.

Writing a Research Report

It should be clear from the previous paragraphs that although research and laboratory reports will usually answer Eisenberg's five investigative questions, there is no single correct research report style. The best way to organize a research report is determined by the type of research being carried out, the author's character and aims, and the audience for whom the report is being written. Accordingly, the following guidelines for

Find "model" reports to help guide you through your first report-writing efforts.

report writing will not offer you a recipe for a "perfect" research report. The keys to a good report include well-executed research and the will and skills to communicate the results of your work effectively.

Having acknowledged that there is no single "correct" report-writing style, it is fair to say that over time a common pattern of report presentation has emerged. Reflecting a strategy for answering the five basic investigative questions, through repeated use that structure has become the one that many readers will expect to see. If you are new to report writing, and unless you have been advised otherwise, it may be useful to follow the general pattern outlined in Table 4.1. If you are more experienced and believe there is a more effective way of communicating the results of your work, try out your own strategy. Remember, however, that you are guiding readers through the work: you will have to let them know if you are doing anything they might not expect.

Short research reports and laboratory reports generally have a minimum of seven sections, which are outlined in Table 4.1. Longer reports may add some or all of the extra materials shown in parentheses. It is likely that most of the reports you are asked to write early in your university career will be short. At higher levels, in honours programs and graduate school, or in the professional world, longer reports may be expected.

Depending on the intended recipient of your report, some subset of the elements in Table 4.1 may be appropriate. For example, for a report in a course, you may be able to omit a separate title page to save paper, and a cover letter may not be relevant. However, the report may be intended to serve a more formal purpose; for example, if you have served as a consultant and are presenting the report to an external organization, there should be a separate title page. Also, attaching a cover letter to the report would show a touch of professionalism.

Although the headings in Table 4.1 point to an order of presentation in the final report, the sections do not have to be written strictly in that order. Indeed, you may find it useful to follow Woodford's advice (in Booth, 1993, p. 2) to label several sheets of paper Title, Summary, Introduction, Methods, Results, and so on, and use these to jot down notes as you work through the project. Then begin your report by writing the easiest section (often the section describing the methods used).

You are not required to write your report in the same order it will be presented.

The following pages outline the form and function of the common components of research reports. They also discuss some of the characteristics that contribute most to effective research presentations in an academic setting. Those same characteristics form the basis of the assessment form for research reports and laboratory reports found at the end of the chapter.

Table 4.1 ◉ Sections of a report*

(Cover letter, if appropriate)
Title page
(Abstract or executive summary)
(Table of contents)
(Acknowledgements [sometimes placed immediately before the List of References])
Introduction
Materials and methods
Results
Discussion
Conclusion
(Recommendations)
List of References
(Appendices)

* Items shown in parentheses are applicable to longer reports and are often omitted from shorter reports.

Preliminary material

Title page

The characteristics of good titles and the elements to include on a title page were covered in Chapter 1. In a few words, a well-selected title can tell the reader much about a report's contents. Cargill and O'Connor (2009) give an example of an effective title for a research article: "Bird use of rice field strips of varying width in the Kanto Plain of central Japan." Through careful construction, this succinct title provides several pieces of information about the research and raises several questions in readers' minds. (e.g., Why was the width of field strips an important variable? How did the researchers measure bird use?) From the title alone, readers may be drawn into reading the article to satisfy their curiosity regarding the answers to those questions.

Cover letter

When a report has been commissioned by an organization or company, as often happens, a cover letter should accompany the report. This letter should be customized to the organization to whom the report is being sent. It usually does the following:

- Explains the purpose of the letter (e.g., "Enclosed is the final report on wetland management issues in eastern New Brunswick that was commissioned by your organization.");

- States the main finding of the report and any other vital issues likely to be relevant;
- Acknowledges any significant assistance received from the organization (e.g., "We are indebted to the Eastern New Brunswick Wetland Lovers Group for allowing us access to their extensive photographic records.");
- Offers thanks for the opportunity to conduct the research (e.g., "We would like to thank the Department of Environment for engaging us to conduct this research.").

(after Mohan, McGregor, and Strano, 1992, p. 227)

Abstract or executive summary

Other than the title, this is the section of the report most likely to be read. You must, therefore, make sure that it is easy to understand. An **abstract** is a short, coherent statement, intelligible on its own, that gives concise answers to each of the five investigative questions outlined in Box 4.1.

> ***Many readers decide from the title and abstract whether or not to read a report; make sure they are clear and accurate.***

Abstracts are short (usually 100–300 words) and are designed to be read by people who may not initially intend to read the whole report. After reading a good abstract, however, they may be intrigued and change their intention! Abstracts are in paragraph form, not point form. All information contained in the abstract must be discussed in greater detail within the main report. Do not write an abstract as if it were the alluring back-cover blurb of a mystery novel. Let your readers know what your research is about and **state** the main findings—do not leave them in suspense. Although the abstract is placed at the beginning of your report, it will usually be the last section you write.

There are two main types of abstract, *informative* and *descriptive* (Goldbort, 2006), although some combine features of both. An *informative* abstract (Box 4.2) summarizes primary research and offers a concise summary of the paper's content, including aims, methods, results, and conclusions.

A *descriptive* or *indicative* abstract (Box 4.3) outlines the contents of a paper, report, or book but does not recount details of the work as an informative abstract does. It "functions primarily to tell readers the kinds of information an article contains, focusing on the research problem" (Goldbort, 2006, p. 99). It is more suitable than an informative abstract for review articles in which secondary, not primary, research is presented.

Abstracts usually consist of a single paragraph, although long abstracts may require more. They do not usually contain figures, tables, or formulae; nor should they normally refer to other works. If you do refer to specific

Box 4.2 ◉ Examples of informative abstracts

Example 1

Masuda, J.R., Zupancic, T., Poland, B. and Cole, D.C. (2008). Environmental health and vulnerable populations in Canada: Mapping an integrated equity-focused research agenda. *Canadian Geographer, 52*(4), 427–450.

Abstract

The uneven distribution of environmental hazards across space and in vulnerable populations reflects underlying societal inequities. Fragmented research has led to gaps in comprehensive understanding of and action on environmental health inequities in Canada and there is a need to gain a better picture of the research landscape in order to integrate future research. This paper provides an initial assessment of the state of the environmental health research field as specifically focused on vulnerable populations in Canada. We present a meta-narrative literature review to identify under-integrated areas of knowledge across disciplinary fields. Through systematic searching and categorization, we assess the abstracts of a total of 308 studies focused on the past 30 years of Canadian environmental health inequity research in order to describe temporal, geographical, contextual, and epistemological patterns.

The results reveal that there has been significant growth in Canadian research documenting the uneven distributions and impacts of environmental hazards across locations and populations since the 1990s, but its focus has been uneven. Notably, there is a lack of research aimed at integrating evidence-based and policy-relevant evaluation of environmental health inequities and how they are created and sustained. Areas for future research are recommended, including more interdisciplinary, multi-method, and preventive approaches to resolve the environmental burden placed on vulnerable populations and to promote environmental health equity.

Example 2

Zimble, D.A., Evans, D.L., Parker, R.C., Grado, S.C., Carlson, G.C., and Gerard, P.D. (2003). Characterizing vertical forest structure using small-footprint airborne LiDAR. *Remote Sensing of Environment, 87*(2–3), 171–182.

Abstract

Characterization of forest attributes at fine scales is necessary to manage terrestrial resources in a manner that replicates, as closely as possible, natural ecological conditions. In forested ecosystems, management decisions are driven by variables such as forest composition, forest structure (both vertical and horizontal), and other ancillary data (i.e., topography, soils, slope, aspect, and disturbance regime dynamics). Vertical forest structure is difficult to quantify and yet is an important component in the decision-making process. This study investigated the use of light detection and ranging (LiDAR) data for classifying this attribute at landscape scales for inclusion into decision-support systems.

Continued

Analysis of field-derived tree height variance demonstrated that this metric could distinguish between two classes of vertical forest structure. Analysis of LiDAR-derived tree height variance demonstrated that differences between single-story and multi-story vertical structural classes could be detected. Landscape-scale classification of the two structure classes was 97 per cent accurate. This study suggested that within forest types of the Intermountain West region of the United States, LiDAR-derived tree heights could be useful in the detection of differences in the continuous, non-thematic nature of vertical forest structure with acceptable accuracies.

Box 4.3 Examples of descriptive abstracts

Example 1

Marks, R.A. (2009). The urban in fragile, uncertain, neoliberal times: Towards new geographies of social justice? *Canadian Geographer*, *53*(3), 345–356.

Abstract

Canadian cities are at a crossroads. The neoliberalization of governance at multiple scales, inadequate re-investment in urban infrastructure, increasing reliance on continental and international trade, and the restructuring of the space economy have combined to weaken Canada's cities just as the global economic system is undergoing transformation. Canadian urban geographic scholarship has much to offer under current conditions, and is already making significant contributions in key areas. In particular, research on what might be called the contours and impacts of urban restructuring and the neoliberal city, immigration and cities of difference, and urban environmental justice show much promise and are likely to define the core of Canadian urban geography into the future.

Example 2

Thompson, P., and Strohm, L.A. (1996). Trade and environmental quality: A review of the evidence. *Journal of Environment & Development*, *5*(4), 363–388.

Abstract

At the heart of the current debate between environmentalists and the trade community lie differences of opinion about the impact that trade liberalization has on environmental quality. This article reviews economists' evidence on the physical trade–environment link. Contrary to interpretations from the trade community, the evidence lends no support to claims that positive income effects from economic expansion outweigh negative scale effects. On the other hand, the evidence does not support environmentalists' arguments that trade is a significant source of environmental degradation, nor that trade liberalization has induced significant international migration of dirty industries.

works they must, of course, be cited and included in the list of references associated with the full report.

It is customary that issues included or discussed in the body of the report are presented in the present tense (e.g., "This report describes the nature of chemical weathering on . . ."; "The report concludes that . . ."), whereas what the author did and found is written in the past tense (for example, "It was discovered that . . ."; "Moreover, weathering had the effect of . . .").

Acknowledgements

If you have received valuable assistance and support from individuals or organizations in undertaking the research or in the preparation of the report, they should be acknowledged. As a general rule you should thank those people who genuinely supported or helped with aspects of the work, such as providing funding, granting access to restricted areas or materials, proofreading, preparing illustrative material, solving statistical or computing problems, solving technical challenges, or taking photographs.

Table of contents

This should accurately and fully list *all* headings and subheadings used in the report with the corresponding page numbers. The table of contents usually occupies its own page and must be organized and spaced carefully. Make sure that the numbering system used in the table of contents is the same as that used in the body of the report.

After the table of contents, and on separate pages, are lists of figures and tables. Each of these lists contains, for each figure or table, its number, title, and the page on which it is located. If your report uses many abbreviations, **acronyms**, or symbols, provide a separate list of these, too, but make sure that you also **define** each one fully when it first appears in the text.

Page numbers in the table of contents and lists of illustrative material should be aligned vertically to produce a clean appearance. In word-processing programs, using right-justified tabs rather than filling up each line with spaces or dots will create a column of page numbers with a *perfectly* aligned edge rather than a ragged or wavy edge.

Introduction

When you write your introduction, imagine that your readers are unfamiliar with your work. The introduction will let them know why this report is important and what exactly it is about (Box 4.4), but it should not include any data or conclusions from your study. When the readers know what you are going

Box 4.4 The introduction of a report

The introduction of a report answers the following questions:

- What question was being asked? (If relevant, state your research **hypothesis**.)
- What did you hope to learn from this research?
- Why is this research important? (What is the social or scientific, personal, or disciplinary significance of the work? This usually requires a literature review.)

to discuss, they will be better able to grasp the significance of the material you present in the rest of your report.

Use the introduction to persuade your audience to read the rest of your report.

When your readers have finished reading the introduction, they should know exactly what the study was about, what you hoped to achieve from it, and why it is significant. If they have also been inspired to read the rest of the report, so much the better!

Literature review

As part of your introduction, or soon after, you may need to write a **literature review** to provide the background to, and justification for, your research. A literature review helps the reader understand the context in which you set up and conducted your research. It is sometimes presented as a separate section of the report, after the introduction and before the discussion of materials and methods. A literature review is a comprehensive, but pithy and critical, summary of publications and reports related to your research. It should discuss significant other works written in the area and make clear your assessment of those works.

Note that, in this context, to be *critical* (and to perform **critical reading** or to provide *criticism*) does not demand that only negative opinions be expressed. Rather, it refers to taking an objective approach, where you ask questions about other work with an open mind rather than from a predetermined position. Questions to ask for a literature review include these: Do you agree with the author? What are the work's strengths and weaknesses? What questions have been left unanswered? How does the work relate to the topic your report discusses? (Cottrell, 2003, p. 210).

The literature review serves a number of functions (Deakin University Library, undated):

- It prevents you from "reinventing the wheel" (that is, repeating the earlier work of others).
- It identifies gaps in the literature and potential research areas.
- It increases your breadth of knowledge in the field and highlights information, ideas, and methods that might be relevant to your project.
- It identifies other people working in the same area.
- It puts your work into intellectual and practical perspective by revealing ways in which it may contribute to, fit in with, or differ from available work on the subject.

A good literature review makes clear the relationships between your work and work that has been done in the area before.

Starting a literature review can be difficult. Chapter 2 lists the range of sources you should read to learn about your topic. Draw from the information in those sources to get some sense of the history of your research topic and of the most important authors, texts, and articles. You may also find it helpful to consult a database, such as the Social Sciences Citation Index (SSCI) or the Science Citation Index (SCI), to **trace** the intellectual genealogy of key sources. (Ask your librarian about this helpful technique.)

Two pieces of advice about references mentioned in Chapter 1 must be repeated: (1) keep track of the bibliographic details of references as you collect them; and (2) insert citations as you write. See Chapter 6 for information about bibliographic-management software programs for storing reference details. These programs can transfer details directly from online databases, thereby eliminating the need to write down or type out reference information.

Once you have gathered and read the relevant resources, you can begin writing the literature review. As a rule of thumb, a good literature review might normally discuss the truly significant publications in the field, notable books and articles produced on the broad subject in the past several years, and the most relevant material on your specific research area. As noted above, the review should give the reader an understanding of the conceptual and disciplinary origins and significance of your study. A literature review requires careful writing, as Reaburn (quoted in Central Queensland University Library, 2000, unpaginated) observes:

> Students will get a pile of articles and will regurgitate what article one said, what article two said. I can't emphasize enough, [that] a well written literature review must evaluate all the literature, must speak generally, with general concepts they have been able to lift from all the articles, and they

must be able to evaluate and critically analyze each one, then link and make a flow of ideas. Rather than separate little boxes, each box representing an article, make a flow of ideas, generalize and use specifics from one or two articles to back up a statement.

Your literature review should not be a string of quotations from other authors' work. Do not make the mistake of trying to list and summarize all the material published in your area of work. Rather, organize the review into sections that present themes and trends related to your research. Integrate all the little pieces of knowledge you have found in your reading into a coherent whole. Your completed literature review should be a critical analysis of earlier work set out in such a way that it becomes evident to the reader why your research work was conducted.

Check to see that your literature review tells a coherent story about the development of research relevant to your topic.

Look at examples of good literature reviews in your area of interest to gain a sense of how they are written. Short reviews are included as part of research papers in virtually every good academic journal. Have a look, for instance, at the reviews by Lucas, Munroe et al., and Pigozzi in a single issue of *The Professional Geographer*.

Lucas, S. (2004). The images used to "sell" and represent retirement communities. *Professional Geographer, 56*(4), 449–459. (Lucas confines her review to pages 450–453 in particular.)

Munroe, D.K., Southworth, J., and Tucker, C.M. (2004). Modeling spatially and temporally complex land-cover change: The case of Western Honduras. *Professional Geographer, 56*(4), 544–559. (The literature review in this paper is brief and confined largely to page 545.)

Pigozzi, B.W. (2004). A hierarchy of spatial marginality through spatial filtering. *Professional Geographer, 56*(4), 460–470. (Pigozzi's review is set out in the section entitled "Introduction and context.")

Materials and methods

You should give a precise and concise **account** of the materials and methods used to conduct the study and why you chose them. Let your reader know exactly *how* you did the study and *where* you got your data. A good description of materials and methods should enable readers to replicate the investigative procedure even if they have no source of information about your study other than your report. In qualitative studies, however, replication of procedure is unlikely to lead to identical results. Instead, the outcome may be results that corroborate, substantiate, or, indeed, refute those achieved in the initial study.

If your readers had no source of information other than your report, could they repeat your study?

Depending on the specific character of the research, the methods section of a report comprises up to three parts, which may be written as a single section or presented under separate subheadings, such as sampling and subjects, equipment and materials, and procedures.

Sampling and subjects

An important part of the materials and methods section of a report is a statement of *how* and *why* you chose some particular place, group of people, or object to be the focus of your study. For example, if your research concerns people's fears and the implications those fears have for the use of urban space, why have you chosen to confine your study to some specific suburb of one North American city? Having limited the study to that location, why and how did you choose a small group of people to interview from the much larger local population? Or in an examination of avalanche hazards in British Columbia's mountains, why did you limit your study to one particular mountain range? There are normally very practical reasons for making such choices.

In the sampling or subjects section of your report, your reader will appreciate answers to the following questions:

- *Whom, or what* specific group, place, or object, have you chosen to study? While you may have already stated in the introduction the general nature of the subject group, you now need to give the precise details.
- *Why* did you make that choice? Why did you choose a particular group of people to interview? On what basis was the sample size or study location chosen?
- *How* did you select your sample to ensure it was representative of the population being studied? That is, what specific sampling technique did you employ (e.g., random selection, stratification, clustering)?
- *What* are the limitations and shortcomings of the data or sources? These may have been unavoidable, but they may affect the generality of conclusions that can be drawn from the results.

In addition to answering the preceding questions, you must also provide a general description of the study site. Include a map of the study area. Photographs may also be helpful. Don't hesitate to illustrate a report with supplementary materials.

Equipment and materials

Give a short description of any special equipment or materials used in your study. For example, describe briefly any experimental equipment or questionnaires used in your work. In more advanced studies you may also be expected to include, in **parentheses**, the name and address of equipment manufacturers. Do not hesitate to use diagrams and photographs in your description of the equipment.

Procedures

This section contains specific *details* about how you collected data and which methods you used to interpret the findings. For example, if your study was based on a questionnaire survey or an experimental procedure, you need to tell your readers about the process of administering the questionnaire, or about the experiment, in enough detail to allow them to replicate your procedures. What were the response levels? What statistical tests did you decide to use? Moreover, you must *justify* your selection of data collection and statistical procedures in this section. Why did you choose one method over others? Give references to support your selection.

What are the advantages and disadvantages of the procedure you selected, and how did you overcome any problems you encountered? You might consider revisiting this last question in your discussion section of your report as to how it may impact your conclusions. Where appropriate (e.g., in research involving human or animal subjects), your discussion of procedural issues should also explain how the work satisfied any relevant guidelines for ethical conduct.

Results

The results section of a research report is a dispassionate, factual account of the findings. It outlines what occurred or what you observed. In this section, you should state clearly whether any hypotheses you made can be accepted or rejected, but, in general, do not discuss the significance of those results here. You will cover that in the discussion section of the report. For example, you may have conducted a study that suggests that deer living on a small island will starve to death unless something is done to control their population. You would save your discussion of options for solving that problem for a later section of the report.

Although it is not customary to present conclusions and interpretations in the results section, in some qualitative and laboratory reports the results are combined with an *interpretive* discussion to help the reader understand them. If you have any doubt about which way is best, ask your professor.

A key to the effective presentation of results is to make them as comprehensible to your readers as possible. To this end, it may be useful to begin your discussion of results with a brief overview of the material that is to follow before elaborating on it. You might also consider presenting your results in the chronological order in which you discovered them.

> ***Results are important, but save enough time, space, and energy to interpret them.***

Use maps, figures, tables, and written statements creatively to summarize and convey the most important information emerging from the study. If you have provided results in figures and tables, do not repeat the data in the text. Emphasize only the most important observations. You should place figures and tables close to where they are mentioned in the text without interrupting the flow of the text. However, if you have particularly detailed and lengthy data lists or figures that supplement the report's content, these are better placed in an appendix.

The results section will also often contain a series of subheadings. These usually reflect subdivisions within the material being discussed, but sometimes they refer to matters of method. In general, however, you should try to avoid splitting up the results section on the basis of methods, since doing so may suggest that you are "allowing the methods rather than the issues to shape the problem" (Hodge, 1994b, p. 2).

Discussion and conclusion

> *I am appalled by . . . papers that describe most minutely what experiments were done, and how, but with no hint of why, or what they mean. Cast thy data upon the waters, the authors seem to think, and they will come back interpreted.*
>
> —Woodford, 1967, p. 744

The discussion is the heart of the report. Perhaps not surprisingly, it is also the most difficult to write and, after the title, abstract, and introduction, the most likely to be read thoroughly by your audience. Readers will be looking to see if your work has achieved its stated objectives. So, take particular care when you are writing this part of your report.

The discussion has two fundamental aims:

1. To explain the results of your study. Why do you think the patterns emerged? Or if you were expecting patterns to emerge that did not, why do you think this happened?
2. To explore the significance of the study's findings. What do the findings mean? What new and important matters have been raised?

Compare your results with trends described in the literature. Embed your findings in their larger academic, social, and environmental contexts. Make explicit the ways in which your work fits in with studies conducted by other people and the degree to which it might have broader importance. Indeed, David Hodge (1994b) makes this point:

> Remember that research should never stand alone. It has its foundations in the work of others, and, similarly, it should be part of what others do in future. Help the reader make those connections (p. 3).

The discussion section is where you should interpret and discuss your findings in detail. Use the conclusion to summarize briefly the major points in the discussion and take the reader back to the questions that were laid out in the introduction (see Box 4.4). The concluding sections of the report might also offer suggestions for improvements or variations of the investigative procedure that could be useful for further work in the field: Where do we go from here? Are there other methods or data sets that should be explored? Has the study raised new sets of questions? Any thoughts you have must be *justified*. Many students find it easy to offer suggestions for change, but fewer support their views adequately.

Recommendations

If your report has led you to a position where it is appropriate to suggest particular courses of action or solutions to problems, you may wish to add a section of recommendations. This section could be included in the conclusion or could be a separate section. In some reports, recommendations are placed at the front, following the title page, to emphasize the importance of the findings to the reader. Keep in mind, however, that recommendations should be based on material or evidence presented in the report.

Reference list

For information on formatting the reference list in a research report, see Chapter 6.

Appendices

Material that is not essential to the report's main argument and is too long or too detailed to be included in the main body of the report is placed in an appendix at the end of the report. For example, you might include a copy of the questionnaire you used, background information on your study area, or pertinent data that are too detailed for inclusion in the main text. A different appendix should be used for each type of material. (The appendices should be labelled A, B, C . . . or 1, 2, 3 . . . .) However, your appendix should *not* be a place to put *everything* you collected in relation to your research but for which there was no place in your report. Appendices are usually located after the references.

Don't use appendices as a dumping ground for data.

Written expression and presentation

Language

Some audiences reading research and laboratory reports expect them to be "objective" and dispassionate and written in the third person (e.g., "it was decided" rather than "I decided"). Consider that expectation when writing your report. If you choose to write in the first person, which recognizes that knowledge is created by people, some audiences may be distracted and unconvinced by your apparent "personal bias." Whatever choice you make, remember that simply writing a report in the superficially "objective" third person does not render it any more accurate than a report written in the first person! (For a lengthy discussion of this, see Mansvelt and Berg, 2005.) If you have any questions about the style of language you should use in your report, ask your professor.

Another matter of language that warrants attention is the use of **jargon**. The word *jargon* has two common meanings:

1. Technical terms used
   (a) unnecessarily or
   (b) when clearer terms would suffice;
2. Words or expressions intelligible only to experts in a particular field of study (Friedman and Steinberg, 1989).

Thody (2006) explains that there will be occasions in report writing when you will find it *necessary* to use jargon in the second sense of the word. Using topic-specific terms in the proper context, rather than general terms, can be more effective for clear and straightforward communication. The correct use of terminology can help to demonstrate the depth of your understanding of the topic. But you should never be guilty of using jargon in the more common first sense of the term. Remember, you are writing to communicate ideas to the intended audience as clearly as possible. Use language that allows you to do that.

Make sure your report is coherent and formatted consistently.

Two final points are to check that your report's text and figures "move from the general to the specific . . . for individual sections and for the report as a whole" (Montgomery, 2003, pp. 143–144) and that your report—despite its many sections—reads as a single, integrated document. This is especially important if your report is the product of a group project and sections have been written by different people or teams, or if you wrote sections in a different order than that in which they are presented. Ensuring that your report flows logically may involve the sometimes lengthy tasks of rewriting some sections; reordering elements of the report; and ensuring that all citations, figures, and tables are formatted uniformly.

Presentation

Be sure that your report is set out in an attractive and easily understood style. Care in presentation suggests care in preparation. People tend to be suspicious of an overly "decorated" report that might be trying to deflect attention from its depth and quality. In a professional environment, such as consulting, they may also question the costs of production.

Remember to use a hierarchical structure for headings where appropriate. (See Box 1.2 in Chapter 1.) The uppermost level of headings should identify the main sections of the report, such as Introduction, Methods, Results, Discussion, Conclusions, and List of References (and Appendices, if applicable). For example, the Introduction section (1) may be broken down at the second level into Background (1.1) and Objectives (1.2). In the Methods section (3), you may first describe the Questionnaire (3.1) used in the study, and then summarize the Study Group (3.2).

Box 4.5 shows the aspects of your report that a marker might use to assess your work and assign an overall mark. These items may be used as a guide for improvements during the revision process.

Writing a Laboratory Research Report

A laboratory research report is a particular form of research report. Hence, the preceding advice is applies to it as well. Typically, a lab report recounts dispassionately and accurately the procedures and results of a specific research

Box 4.5 ◉ Research report assessment form

Student Name: **Grade:** **Assessed by:**

This form provides an itemized rating scale for various aspects of a research report. Sections left blank are not relevant to the attached assignment. Some aspects are more important than others, so there is no mathematical formula connecting the scatter of ticks with the final grade for the assignment. Ticks in either of the two boxes left of centre mean that the statement is true to a greater (outer left) or lesser (inner left) extent. The same principle is applied to the right-hand boxes. If you have any questions about the individual scales, comments, final grade, or other aspects of this assignment, please see the assessor indicated above.

Purpose and Significance					
Statement of problem or purpose is clear and unambiguous					Statement of problem or purpose is unclear or ambiguous
Research objectives are outlined precisely					Research objectives are unclear
Significance and context of the research problem is clear					Problem is not set in context
Documentation outlines fully the evolution of the research problem from previous findings					No reference given to earlier works, or references are incorrect
Description of Method					
Most appropriate research method was selected					Research method selected was inappropriate
Sample, case, or study area are appropriate to purpose of inquiry					Sample, cases, or site are unsuitable
Description of study method is complete					Description of study method is inadequate
Quality of Results					
Evidence of extensive primary research					Little or no evidence of primary research
Limitations of sources made clear					Inappropriate sources accepted without question
Relevant results presented with appropriate level of detail					Relevant results omitted or suppressed

Continued

Discussion and Interpretation

No errors of interpretation (e.g., logic, calculation) detected					Many errors of interpretation detected
Limitations of findings are made clear					Limitations of findings are not identified
Discussion connects findings with relevant literature					No connection between findings and other works

Conclusions

Significance of findings is made clear					Little or no significance is identified
Conclusions are based on evidence presented					Little or no connection made between evidence and conclusions
Stated purpose of research was achieved					Little or no contribution to solution of problem or achievement of purpose

Use of Supplementary Material

Illustrative material is used effectively					Illustrative material not used when needed or not discussed in text
Illustrations are presented correctly					Illustrations are presented poorly
Detailed materials are placed in appendices					Excessively detailed findings are given in main body

Written Expression and Presentation

Documentation follows assignment formatting guidelines					Little or no adherence to presentation guidelines
Written expression is clear and correct					Written expression is poor
Report produced carefully					Presentation is sloppy

Sources and Referencing

Enough sources were used					Not enough sources were used
Acknowledgement of sources is adequate					Acknowledgement of sources is inadequate
In-text citation style is correct and consistent					In-text citation style is incorrect or inconsistent
Reference list is presented correctly					Reference list contains errors and inconsistencies

Assessor's comments

experiment. A good report is written so that another researcher could repeat the experiment in exactly the same way as you did (assuming, of course, that you employed correct procedures) and could compare his or her results with yours. You must be both meticulous in outlining your methods and accurate in presenting your results. *Meticulous* does not mean "tedious," however. Try not to overdo the detail. In the context of the experiment you are doing, describe the things that are important. What did you need to know in order to do the experiment? Let your reader know that. One could argue that the best "materials and methods" sections in laboratory reports ought to be written by novices who are less likely than more experienced researchers to make assumptions about readers' understanding of experimental methods.

By convention, laboratory reports follow the same order as research reports (see Table 4.1). Depending on the nature of your experiment, your instructor may not require all of the sections listed. For example, if your report is simply an account of work you undertook during a single laboratory class, there may be no need for an abstract, references, or appendices. However, your professor may be impressed if you take care to relate your day's laboratory work to appropriate reference or lecture material. Clearly, though, if your lab work is conducted over a longer period than a single class, you will have the opportunity to consult relevant reference materials.

Writing a Research Proposal

This section is directed toward readers about to undertake an individual research venture, such as an honours thesis project. Shorter research proposals may be required before embarking on a major term paper. Where appropriate, you may omit or condense some of the sections described below to fit the situation. Also consider whether preparing a research ethics application or environmental impact statement (items discussed at the end of Chapter 2) is required at the outset of a research project.

Purpose

In contrast to course work where the topic is assigned by the professor, individual research requires the student to select the topic and prepare a research plan. Typically, before the actual investigation, experiment, or field study begins, a research proposal is required. This enables a supervisor or committee to determine whether you have acquired a sufficient understanding of the topic and the context in which the research will be conducted. You also need to show that you have developed a solid research plan and have clearly identified your research question or **hypothesis**. Finally, you need to describe the methods you will use so that flaws or shortcomings can be identified and addressed before you begin the research.

> ***A research proposal will be used to assess whether you are prepared to begin a project.***

Characteristics

The length of a research proposal will vary depending on the size of the project. A term paper proposal might require only a page or two to show evidence that sufficient preliminary investigation into the topic has been conducted. For an honours thesis project, two to four pages might be adequate to serve the purpose of describing your proposed research plan without giving more detail than necessary. Each institution normally has guidelines defining the required characteristics of a proposal, such as length. Ask your supervisor if there is a standard protocol to follow.

Common sections of research proposals include the following:

- *Introduction*. Lead the reader into the research topic and briefly summarize the context in which the research will be undertaken. You need to have done at least a preliminary review of relevant literature to understand the context of the research you are proposing. Highlight the most significant publications that have guided the selection of your topic and proposed methods.

- *Description of research question or hypothesis*. Whereas the introduction gives the reader a general idea of the topic, here you must be specific about what you plan to study. If your research fits a hypothesis-testing model, state your hypothesis or hypotheses clearly. After these first two sections the reader should have a clear understanding of what you hope to have achieved when you present your final research report. This does not mean you must try to predict the answer you will find to your question. Obtaining the answer—regardless of what the answer turns out to be—is your objective and the reason for doing the research.
- *Proposed methods*. Give the reader information about how you plan to answer your research question. Will your methods be adequate and appropriate for the task? If there are problematic issues, these can be identified and discussed. Adjusting the methodology now can save considerable time and stress later. In some cases a project that is doomed to fail because of poor methodology can be saved even before it begins. Describe elements of your plan, such as the study group, sampling methods, equipment to be used, analytical procedures, and statistical tests. If you have thought through the project to the end, you should have identified all of the methods you will need to collect, process, and analyze data. Students often focus only on the first parts of data collection without giving sufficient thought to later steps, which can result in not having the appropriate data to process and analyze in order to achieve your research objective.
- *Timeline for research*. Thinking through the research to the end will help give you a sense of what is required for the project to be successful. Be as detailed as you can in creating a list of the key milestones along the way. Use the list to allot an appropriate amount of time to complete the steps. But be careful because a common mistake is to create a list of milestones that is so vague as to be almost useless: conduct literature by the end of one month; finish data collection by the end of another month; complete data analysis by the end of the next month; and write up the research report by the required deadline. A list such as this is an indication that an insufficient amount of detailed thought has gone into the actual steps that will be required.
- *List of references*. All references cited in the proposal need to be included in the list. See Chapter 6 for information on citing sources and preparing a list of references.

Some students worry that the proposal will lock them into a research project that cannot be modified. This is an incorrect view. Research plans often have to be updated and changed as you proceed. Being too rigid and not adapting to the situation can be pointless if the original plan is leading to a dead end. However, do not take this as an invitation to treat the proposal lightly and avoid in-depth forward planning. Even if it will ultimately change later, the research proposal must be specific and detailed. Vague proposals often lead to difficult research projects because problems have not been identified in advance, which can make it challenging to enjoy conducting and reporting on the research.

Research proposals are plans for the work to be undertaken but are not so rigid that they cannot be modified along the way.

Other misconceptions about research proposals involve what should *not* be included. Although you may have had to do *some* preliminary work to test the viability of an idea—work that should be described to help show that your full plan is valid—the proposal is not the place to provide detailed results. If you already have detailed results to report, your proposal is probably being written at the wrong stage in your project (that is, too late). Also, while some background review of the literature must normally be done in order to write the proposal and set the context, an extensive literature review is not required. Conducting a comprehensive literature review is typically a milestone in the research project itself. Restrict the literature review in the proposal to the key references needed to support your proposal and get the project on the way.

Further Reading

Booth, V. (1993). *Communicating in science: Writing a scientific paper and speaking at scientific meetings* (2nd ed.). Cambridge, UK: Cambridge University Press.

- Chapter 1 of Booth's readable and brief book offers helpful advice on writing a scientific paper. Particular attention is devoted to the mechanics and details of scientific presentation.

Day, R.A. (1994). *How to write and publish a scientific paper* (4th ed.). Cambridge, UK: Cambridge University Press.

- Several chapters of this book are devoted to the writing of the various sections of a research report.

Goldbort, R. (2006). *Writing for science*. New Haven, CT: Yale University Press.

- This is a wide-ranging work about writing and communicating in science, from keeping laboratory notes to preparing grant proposals. Chapter 4 is specifically about undergraduate science reports.

Johnson, J. (1997). *The Bedford guide to the research process* (3rd ed.). Boston, MA: Bedford Books.

- This book is strongly recommended for students undertaking independent research projects. It guides the student logically through the research process. Chapter 1, which is particularly useful, discusses the choice of a topic.

Kanare, H.M. (1985). *Writing the laboratory notebook*. Washington, DC: American Chemical Society.

- Chapter 3, "Organizing and writing the notebook," offers detailed advice on keeping a comprehensive laboratory notebook. Other parts of the text discuss matters such as the legal and ethical aspects of note taking, patents, and protection of inventions.

Mohan, T., McGregor, H., and Strano, Z. (1992). *Communicating! Theory and practice* (3rd ed.). Sydney, Australia: Harcourt Brace.

- Chapter 9 offers an overview of report types, functions, format, and style.

Montello, D.R., and Sutton, P.C. (2006). *An introduction to scientific research methods in geography*. Thousand Oaks, CA: Sage.

- This book is suitable as a required textbook in an undergraduate research methods course. The focus is more on various methods that may be used in geographic research than on writing.

Montgomery, S.L. (2003). *The Chicago guide to communicating science*. Chicago: University of Chicago Press.

- Chapter 10, "Technical reports," is a helpful discussion of the context within which reports are written and the form they should take.

Moxley, J.M. (1992). *Publish, don't perish: The scholar's guide to academic writing and publishing*. Westport, CT: Praeger.

- This text contains two chapters that outline different ways of writing reports depending on whether they are based on qualitative or quantitative research.

Robson, C. (2007). *How to do a research project: A guide for undergraduate students*. Malden, MA: Blackwell.

- This book provides a thorough review of the steps in a research project in a succinct but effective style.

University of California—Santa Cruz, University Library. (undated). *Write a literature review.* http://library.ucsc.edu/ref/howto/literaturereview.html (accessed 21 November 2013).

- This website is a helpful and practical resource for those seeking additional guidance on writing a literature review.

University of Wisconsin—Madison Writing Center. (2012). *Review of literature.* http://www.wisc.edu/writing/Handbook/ReviewofLiterature.html (accessed 18 November 2013).

- This is another helpful, practical resource for those seeking additional guidance on writing a literature review.

5 Writing an Annotated Bibliography, Summary, or Review

I was so long writing my review that I never got around to reading the book.

—Groucho Marx

Some books are to be tasted, others to be swallowed and some few to be chewed and digested.

—Francis Bacon

Key Topics

- Preparing an annotated bibliography
- Writing a summary or précis
- Writing a review

Your academic endeavours will often require you to summarize and make sense of the works of other people. This chapter offers some advice on writing annotated bibliographies, summaries, and reviews of books, articles, and websites—exercises that specifically require you to interpret and abridge longer pieces of work comprehensibly. Where appropriate, the discussion sets out the criteria that readers and assessors usually consider in evaluating this kind of work. You will find assessment criteria for different types of summary assignments in Boxes 5.3 and 5.4.

Preparing an Annotated Bibliography

An **annotated bibliography** is a list of reference materials, such as books, articles, and websites, in which you give the author, title, and publication details for each item (as in a **bibliography**) *along with* a short **summary** and **critique** of that item. The summary and critique section might be up to 150 words long. Annotated bibliographies customarily list the items in alphabetical order (by author's surname). Some annotated bibliographies, however, are written in the form of a very short essay that quickly and concisely offers the same bibliographic material and critique but in a more literary style than an annotated list.

Box 5.1 gives examples of items reproduced from annotated bibliographies. An online example of a complete annotated bibliography (with sufficient entries to warrant an index) is the one prepared by Cottingham et al. (2002).

Box 5.1 ● Examples of items in annotated bibliographies

Example 1

Northcott, M.S. (2007). *A moral climate: The ethics of global warming*. London: Darton, Longman and Todd.

Response to the challenge of global warming requires learning to put the common good ahead of selfish interests, weaving together the physical climate and the moral climate. Relieving climate change opens opportunities for solving other problems: world poverty, the rich–poor divide, the overuse of resources, and the appreciation and conservation of human creation.

Source: Callicott and Froderman (2008).

Example 2

Sturgeon, N. (2009). *Environmentalism in popular culture: Gender, race, sexuality, and the politics of the natural*. Tucson: University of Arizona Press.

This is the first book to employ a global feminist environmental justice analysis to focus on how racial inequality, gendered patterns of work, and heteronormative ideas about the family relate to environmental questions. Beginning in the late 1980s and moving to the present day, Sturgeon unpacks a variety of cultural tropes, including ideas about Mother Nature, the purity of the natural, and the allegedly close relationships of indigenous people with the natural world. She investigates the persistence of the "myth of the frontier" and its extension to the frontier of space exploration. She ponders the popularity (and occasional controversy) of penguins (and penguin family values) and questions the assumptions about human warfare as "natural."

Sturgeon illustrates the myriad and insidious ways in which American popular culture depicts social inequalities as "natural" and how our images of "nature" interfere with creating solutions to environmental problems that are just and fair for all. Why is it, she wonders, that environmentalist messages in popular culture so often "naturalize" themes of heroic male violence, suburban nuclear family structures, and US dominance in the world? And what do these patterns of thought mean for how we envision environmental solutions, like "green businesses," recycling programs, and the protection of threatened species?

Source: Reed (2011).

Continued

Example 3

Bullard, R.D. (1990). *Dumping in Dixie: Race, class, and environmental equity.* Boulder, CO: Westview Press.

Dumping in Dixie is an in-depth study of environmental racism in black communities in the South. Bullard explores the barriers to environmental and social justice experienced by blacks and the factors that contribute to the conflicts, disparities, and the resultant growing militancy. He provides case studies of strategies used by grassroots groups who wanted to take back their neighborhoods in Houston's Northwood Manor neighborhood; West Dallas, Texas; Institute, West Virginia; Alsen, Louisiana; and Emelle-Sumter County, Alabama.

In these predominantly black communities, grassroots organizing was carried out to protest against landfills, incinerators, toxic waste, chemical industries, salvage yards, and garbage dumps. Strategies included demonstrations, public hearings, lawsuits, the election of supporters to state and local offices, meetings with company representatives, and other approaches designed to bring public awareness and accountability. Bullard offers action strategies and recommendations for greater mobilization and consensus building for the ensuing environmental equity struggles of the 1990s.

Source: Weintraub (1994).

What is the purpose of an annotated bibliography?

Annotated bibliographies provide people in a particular field of inquiry with some commentary on resources available in that field. They discuss the content, relevance, and quality of a body of material (Engle, 2013). Annotated bibliographies may also give a newcomer to a body of work an insightful introduction to material available. Thus, your professor might ask you to write an annotated bibliography to make you familiar with some of the literature in your discipline. A student beginning a research project would benefit from preparing an annotated bibliography to help summarize key information in the field of study.

What is the reader looking for in an annotated bibliography?

Although the content may vary depending on the purpose of the annotated bibliography, readers typically expect to see the three sets of information, described in Box 5.2—details, summary, and critique—although many annotated bibliographies exclude the element of critique.

Information in the remainder of this chapter can be applied to the preparation of an annotated bibliography. The next section discusses writing a summary, and the final section concerns writing a review, which includes

Box 5.2 Key information in an annotated bibliography

- *Details.* Full bibliographic details about the source material (see Chapter 6).
- *Summary.* A clear indication of the content and argument(s) of the material. Consider including, for example, material on the author's aim in writing the piece; the intended audience; the author's claim to authority; and the main arguments used to support points.
- *Critique.* Critical comment on the merits and weaknesses of the work or on its contribution to the field of study. Some of the things you might also consider evaluating are the suitability of the piece to its intended audience; whether the work is up-to-date; and, especially, its engagement with other important literature in the field.

elements of analysis and evaluation (that is, critiquing skills). In an annotated bibliography, you would not write nearly as much as for a full summary or review of a single work, but the discussions that follow may increase your understanding of these tasks.

Writing a Summary or Précis

A summary (sometimes called a **précis**) restates the essential contents of a piece of writing in far fewer words than the original text. Shortening the text is achieved by presenting the main ideas in an alternative wording and leaving out most examples and minor points.

Your summary must accurately *re-present* the text in condensed form. It should be a scaled-down version of the original text. Unlike a review, a summary does *not* analyze the issues raised. It is *not* evaluative. There is no need to mention your reaction to the author's ideas.

Two important ingredients of a summary are *brevity* and *clarity*. Let your reader know, in as few words as possible, what the summarized text is about. Do not write a summary that is as long as the original work. Give your readers enough information to understand what the text is about but not so much that they might just as well have read the original.

A good summary spares the details and focuses on key matters.

This discussion leads us to one of the most common shortcomings of a summary: the failure to clearly identify the original author's *main argument*. Imagine that someone has asked you to tell them about a movie you have seen. One of the most important things they will want to know, and in relatively few words, will be *the essential elements of the plot*. For example, readers probably do not want to know the names of Cinderella's evil sisters, the colour of her ball gown, or the temperament of the horses. Instead, they want to know the essence of the story that brought Cinderella together with Prince Charming.

Once you have written your summary, reread it with the following question in mind: "Could I read this piece aloud to the author in the honest belief that it accurately summarizes his or her work?" If the answer is no, modify your work. You might also ask yourself this question: "Would someone who has not read the original text have a good sense of what that text is about after reading this summary?" Again, if the answer is no or if some parts have not been summarized, you have revisions to make.

What is the reader of a summary looking for?

The criteria in the assessment form in Box 5.3 can be used as guidelines for writing your summary or précis.

Full bibliographic details of the work

At the start, you should provide the reader with a full bibliographic reference to the work you are summarizing. Consult Chapter 6 for detailed information on correct referencing techniques.

Clearly stated subject matter and purpose

After you have compiled the references, you should provide a short statement informing your readers what the reviewed work is about. Then, having stated *what* the work is about, you need to let readers know precisely what the *purpose* of the work is. The distinction between the subject matter of a work and its purpose is illustrated by the following introductory passage from Ley and Bourne's edited collection on the social geography of Canadian cities:

> This is a book *about* the places, the people and the practices that together comprise the social geography of Canadian cities. Its *purpose* is both to describe and to interpret something of the increasingly complex social characteristics of these cities and the diversity of living environments and lived experiences that they provide (Ley and Bourne, 1993, p. 3; emphasis added).

In his **preface** to *Recent America*, Dewey Grantham also distinguishes between the subject matter (first two sentences) and the purpose (third sentence) of his work:

Box 5.3 ◉ Summary (or précis) assessment form

Student Name: **Grade:** **Assessed by:**

This form provides an itemized rating scale of various aspects of a summary or précis. Sections left blank are not relevant to the attached assignment. Some aspects are more important than others, so there is no mathematical formula connecting the scatter of ticks with the final grade for the assignment. Ticks in either of the two boxes left of centre mean that the statement is true to a greater (outer left) or lesser (inner left) extent. The same principle is applied to the right-hand boxes. If you have any questions about the individual scales, comments, final grade, or other aspects of this assignment, please see the assessor indicated above.

Description					
Full bibliographic details of the text are provided					Insufficient bibliographic details are provided
Text's subject matter is identified clearly					Text's subject matter is defined poorly or inadequately
Purpose of the work is identified clearly					Purpose of the work is unclear or not stated
Emphases in the summary match emphases in original work					Little or no correspondence with work's emphases
Order of presentation in summary matches that of original work					Little or no correspondence with work's order of presentation
Key evidence supporting the original author's claims is outlined fully					Little or no reference is made to original work's evidence
Summary is written in student's own words					Summary is constructed largely from quotes

Written expression and presentation					
Writing is fluent					Writing is clumsy, verbose, or repetitive
Sentences are grammatically correct					Sentences are not grammatically correct
Punctuation is correct					Punctuation is poor

Continued

Spelling is correct throughout					Spelling is poor
Work is legible and well laid out					Work is untidy or difficult to read
Length is reasonable					Work is too short or long
In-text citation style is correct and consistent					In-text citation style is incorrect or inconsistent
Reference list is presented correctly					Reference list contains errors or inconsistencies

Assessor's comments

> *Recent America* seeks to provide a relatively brief but comprehensive survey of the American experience since 1945. The emphasis is on national politics and national affairs, including international issues and diplomacy, but some attention is given to economic, social, and cultural trends. I hope that this volume will serve as a useful introduction to a fascinating historical epoch (Grantham, 1987, p. x).

Be sure to distinguish between the subject matter and purpose of the work you are summarizing.

You must think carefully about the distinction between a work's subject matter and its purpose. Usually a book or article will discuss some topic or example in order to make or illustrate a particular point or to investigate a specific theme. For example, a book about the transmission of inherited housing wealth to women in Quebec may actually be trying to contribute to broader discussions of Canada's political economy. Or an article about a technique for satellite image analysis may have been written for the purpose of detecting the effects of a disease in stands of trees. You need to be sure you have not confused the subject matter and the purpose.

Summary of the original work

The summary is written in your own words and should match or reflect the original work. For example, in the summary you should give the same relative emphasis to each area as the authors of the original work do (Northey et al., 2012). If, for example, two-thirds of a paper on monitoring water pollution from industrial sites in southern Ontario is devoted to a discussion of the legalities of obtaining the water samples, your summary should devote a similar proportion of its attention to that issue. This helps to give your reader an accurate view of the original text.

Just as you should give the same emphasis to each section as the author of the original work did, in your summary you should also follow its order of presentation and chain of argument (Northey et al., 2012). Make sure you have presented enough material for a reader to be able to follow the logic of each important argument. You will not be able to provide every detail. Present only the most important connections.

If you are reviewing a website, however, this advice on following the order of presentation may be redundant given the non-linear structure of some online resources. Consider instead trying to follow the structure of the site indicated by the main links on the site's home page.

Summarizing a website presents unique challenges.

Although you want to reflect the work of the original author, be careful to write the summary primarily in your own words. You may, of course, elucidate some points with quotations and paraphrases from the original source. But do not construct your summary predominantly as a collection of direct quotations or paraphrases from the work you are summarizing.

Key evidence supporting the author's claims

You should briefly mention the key evidence provided by the author to support his or her arguments. There is no need to recount all the data or evidence offered by the author. Instead, refer to the material that you found to be the most compelling and convincing.

Writing a Review

A review is an honest, concise, and thoughtful description, analysis, and evaluation of some work, such as a book, article, research report, or website. Reviews differ from summary reports in that they include evaluation elements in addition to analysis. Reviews have an important place in the professional and academic world (Lindholm-Romantschuk, 1998), and for that reason you will see review sections (most commonly for new monographs) in almost every major journal of environmental science or geography. Reviews let people know of the existence of a particular work and point out its significance. They also warn prospective users about errors and deficiencies. However, because so many new publications and online resources are available, we need to be selective about what we read.

Professors usually ask you to write reviews for one or more of the following reasons:

- To familiarize yourself with a significant piece of work in the field;
- To evaluate the importance of a work to the discipline you are studying;
- To practise critical analysis.

People who read reviews, including those marking your review, want *honest* and *fair* comments on the following aspects of the work:

- *Description*: what the reviewed item is about;
- *Analysis*: details of its strengths and weaknesses;
- *Evaluation*: its contribution to the discipline.

A review should be interesting as well as informative. A summary of the work generally precedes an analysis, but there are no rules for the order in which you should present the elements of a review. Instead, you should set out the material in a manner that is both comprehensive and interesting. Before writing your own review, you should read a few reviews in academic journals in your discipline to see how they have been structured.

Reviews are vital, critical guides to new resources.

Box 5.4 shows a sample of criteria for assessing a review of a published work. The next few pages of guidelines and advice will help you to meet these criteria.

Summary: What is this work about?

Description of the work

Description or summary—described in the section above—is an important part of a review. As with a summary, you must provide a full and correctly formatted reference to the work under review so that others may consult or acquire it. You should imagine that your readers have not read the work you are discussing and that their only knowledge of it will come from your review. Give them a comprehensive but concise outline of the content and character of

Box 5.4 ◉ Assessment form for a review

Student Name: **Grade:** **Assessed by:**

This form provides an itemized rating scale of various aspects of a review of a published work. Sections left blank are not relevant to the attached assignment. Some aspects are more important than others, so there is no mathematical formula connecting the scatter of ticks with the final grade for the assignment. Ticks in either of the two boxes left of centre mean that the statement is true to a greater (outer left) or lesser (inner left) extent. The same principle is applied to the right-hand boxes. If you have any questions about the individual scales, comments, final grade, or other aspects of this assignment, please see the assessor indicated above.

Description

Full bibliographic details of the work are provided					Insufficient bibliographic details are provided
Sufficient details of author's background are given					No details of author's background are given
Work's subject matter is identified clearly					Work's subject matter is identified poorly or inadequately
Purpose of the work is identified clearly					Purpose of work is unclear or not stated
Author's conceptual framework is identified correctly					Little or no attempt is made to identify conceptual framework
Summary of work's content is succinct					Summary of work's content is excessive or inadequate
Intended readership is identified accurately					Intended readership is not identified

Analysis

Work's contribution to understanding of the world or the discipline is identified clearly					Little or no reference to work's contribution is provided
Clear statement on achievement of work's aims is given					Work's aims are explained incorrectly or not at all
Work's academic or professional functions are stated clearly					Work's functions are stated incorrectly or not at all

Continued

Work's organization is commented on fairly					Little or no comment is given on work's organization
Work's evidence is evaluated critically					Little or no comment is given on work's evidence
Work's references are evaluated critically					Little or no comment is given on work's references
Style and tone of presentation are evaluated critically					Little or no comment is provided on style and tone
Quality of supplementary material (e.g., tables, maps, plates) is reviewed competently					Little or no comment is provided on supplementary material
Other deficiencies or strengths in work are identified correctly and fairly					Other evident weaknesses or strengths of work are not identified

Evaluation

Work is compared usefully with others in the field					Little or no effort is made to compare work with others in the field
Valid recommendation is made about the value of reading the work					No recommendation is made about the value of reading the work, or recommendation is inconsistent with earlier comments

Written expression, references, and presentation of the review

Various sections of review are of appropriate length					Major imbalances in sections of review are evident
Writing is fluent					Writing is clumsy, verbose, or repetitive
Sentences are grammatically correct					Sentences are not grammatically correct
Punctuation is correct					Punctuation is poor
Spelling is correct throughout					Spelling is poor
Text is legible and well laid out					Text is untidy or difficult to read

Length is reasonable					Text is too short or long
In-text citation style is correct and consistent					In-text citation style is incorrect or inconsistent
Reference list is presented correctly					Reference list contains errors or inconsistencies

Assessor's comments

the work. Remember to be clear in differentiating between the subject matter and purpose of the work. Do not make the mistake of presenting only a summary without the analytical and evaluative elements that distinguish a review.

You might want to integrate the summary of content with your evaluation of the work, or you might prefer to keep the two separate (Northey et al., 2012). If you are new to review writing, it is usually safer to separate the two. Write the summary first, then the analysis.

Author's background

If you consider it useful and you know of the author's expertise in the area he or she is writing about, give a brief overview of the author's background and reputation (Marius and Page, 2002; Northey et al., 2012). It may also be helpful to find out a little more about an author's affiliation and credentials. You might consider whether the author has written other materials in this area and whether he or she has any practical experience. Readers unfamiliar with the subject will often appreciate some information on the author's areas of expertise, in order to assess their credibility.

Conceptual framework

Books and articles are written from a particular perspective. Each author has a unique way of viewing the world and of arranging his or her observations into some specific and supposedly comprehensible whole. That way of thinking about the world is known as a **conceptual framework**. Imagine it as the work's skeleton, upon which the flesh of words and evidence is supported. As a reviewer, one of your tasks is to expose that skeleton, thereby letting your reader know how the author has interpreted the issues that are discussed. How has the author made sense of that part of the world he or she is discussing?

Try to uncover the intellectual scaffolding that gives the book you are reviewing coherence.

In your review, you might combine a description of the conceptual framework with a critique of it. Does the author make unwarranted assumptions, and are there inconsistencies, flaws, or weaknesses in the intellectual skeleton? For example, an author might argue that massive job losses, associated with the adoption of new labour-saving technologies in industry, have been a feature of capitalism in Canada's manufacturing sector. A reviewer might suggest that this reflects a simplistic view of relations between people and technology under capitalism and might go on to argue that service-sector employment has risen at the same time as industrial job losses have occurred. Moreover, the reviewer might suggest that while some industries have atrophied, others have emerged and grown.

Intended readers

The reader of a review is usually interested in knowing what sort of audience the author of the original work was addressing. In many cases the author will include a statement about the intended readership early in the volume—often in the preface. For example, in the preface to *The Slow Plague*, Peter Gould describes his book as follows:

> . . . one of a series labelled *liber geographicus pro bono publico*—a geographical book for the public good, which sounds just a bit pretentious until we translate it more loosely as "a book for the busy but still curious public" (Gould, 1993, pp. xiii–xiv).

Howitt et al. (1996, p. v) suggest that readers of their edited collection *Resources, Nations and Indigenous People* are likely to include students, Aboriginal organizations, and mining company staff. In the preface to *Introducing Human Geography*, Waitt et al. (2000, p. vii) recognize their readers explicitly: "This textbook introduces first-year university students to contemporary themes and practices in human geography."

Identifying the intended audience of the reviewed work serves at least two important purposes. First, you will be helping your readers decide whether the original work is likely to be of relevance to them. Second, you will be giving yourself an important foundation for writing your critique of the work. For example, from Gould's statement above, it is reasonable to conclude that his book ought to be easy to read, stimulating, and written for an audience without a specialized background in geographical concepts. If it is not, there is an important flaw in the book, that is, a disjunction between the author's intended audience and the work's actual contents or level of information. A book or article intended for experts in the field may legitimately use technical terms and express complex ideas. A version of the same information for children, on the other hand, may be simplified and written using a completely different vocabulary. You should write your review with the relationship between the content and style and the intended audience in mind.

Be sure to identify a work's intended audience and to review it with that audience in mind.

Analysis: What are the strengths and weaknesses of this work?

So far you have let your reader or assessor know some basic descriptive details about the work you are reviewing. Now you need to let everyone know what you consider to be the work's weaknesses *and* its strengths. Try to be balanced in your review even if your overall evaluation of the book is strongly negative or positive (Brozio-Andrews, 2007). If you believe that the material you are reviewing has no significant weaknesses, you should say so. However, you should also point out its specific strengths. As noted by the Writing Center, University of North Carolina at Chapel Hill (2013, unpaginated),

> You're entitled—and sometimes obligated—to voice strong agreement or disagreement. But keep in mind that a bad book takes as long to write as a good one, and every author deserves fair treatment.

In your analysis, you should above all be honest but also fair (Brozio-Andrews, 2007). To do so, explain clearly why you agree or disagree with the author's methods, analysis, or conclusions. In organizing their analysis of a work, many reviewers point out the utility and successes of the book, article, or website first and then move on to point out its deficiencies and errors. You may find that pattern a useful one to follow.

Be fair, be explicit, and be honest when you write a review.

Purpose, function, audience, and organization

Be sure to comment on the work's purpose. Think carefully about the author's objectives and compare those with the content of the work. Do they match one another? You would be derelict in your duty as a reviewer if you stated what the reviewed work's purpose was but failed to say whether or not it had been met.

As well, you should inform your reader of the work's function, including any educational, research, or professional functions. For example, is the book, article, or website likely to be useful to people in the same course as you, to other undergraduate students, or to leaders in the field?

Accentuate the positive. Your analysis will be more helpful if you offset a negative observation with a suggestion that gives the reader an alternative to consider. For example, an author may think that the book's audience ought to be final-year undergraduate students, but you—as an undergraduate student reviewer—believe the book would better serve a lower-level audience. Rather than simply stating that the work is unsuitable for its intended audience, let your reader know which audience the work might best serve.

Try to be helpful in your analysis.

In addition, be sure to state whether or not you believe the reviewed work is well organized. In this regard, think about the ways that the work is divided into chapters or sections or how easy it is to navigate around the website. Do the divisions, chapters, or links advance the work's purpose, or are they obstructive? Do they break up or upset an intellectual trajectory? Support your viewpoint with examples.

Evidence

An author's evidence should be reliable, up to date, drawn from reputable sources (such as official statistics, government reports, international journals, experts in the field, reports from non-governmental agencies, or links to credible and relevant websites), and should support claims made in the work. For research-based material, consider whether the results of the original research would stand up to replication (that is, if the study were done again, is it likely that the results would be the same?) or can be corroborated (that is, are the results of the study substantiated by other related evidence?).

Give a clear assessment of the evidence used in the work. Is the evidence compelling, or not? If, for example, you are skeptical about the sources used to support a discussion of inconsistencies in electoral procedures in US presidential elections, you should let the reader know. Back up your assessment of the evidence with reasons for your conclusion. If you are able to, and if it is necessary, suggest alternative and better sources of evidence.

References

Readers and markers of your review will wish to know if the work reviewed has covered the available and relevant literature to a satisfactory degree. If you are reviewing a website, you might expand your notion of references cited to include links from the site to other relevant websites. If there are major shortcomings in the references acknowledged, it is possible that the work's authors may not be fully aware of material that might have illuminated their work. If you are new to your discipline, you might protest, sometimes quite justifiably, that you cannot offer a meaningful opinion about the strength of the reference material. Nevertheless, you ought to be thinking about this question and answering it where possible.

Style and tone

Among other things, readers of a review may be trying to decide whether to buy, read, or consult the work being reviewed. Therefore, among the questions foremost in the minds of some will be whether the work is written clearly and is interesting to read. Is the writing repetitious? Detailed? Not detailed enough? Boring? Is the style clear? Is it tedious? Full of jargon? Curt? Particularly if you think there are problems with the style of writing, it is appropriate to support your criticisms with a few examples.

If you are reviewing a website, you might also consider commenting on the visual character and quality of the site. Is judicious use made of technical devices, or have distracting electronic wizardry and gadgets stolen the show?

You should also comment on the tone of the work. Let your reader or assessor know if the work is accessible only to experts in the field or would be more suitable as a coffee-table book. Of course, criticism about the work's tone should be written with reference to the intended audience. It would probably be unfair, for instance, to condemn a work on the grounds that it uses technical language if it was written with an audience of experts in mind.

Illustrative material

Many books, articles, and websites make extensive use of illustrative material, such as figures, tables, maps, and plates. In reviewing the work, you should comment on the quality of these materials and their contribution to the author's message. The graphic and tabular material should be relevant, concise, large enough to read, and comprehensible; it should also provide details about the sources.

Other weaknesses or strengths

After a close re-reading of the work, you should be able to recognize any weaknesses and strengths you have not already discussed. Remember: you do not *have* to find things wrong with the work you are reviewing. Being critical does not require you to be negative if it is not warranted. You should not pick out minor problems and suggest that they destroy an entire work. However, if

Support criticisms in your review with evidence.

there are genuine problems, be explicit about what they are and give examples where possible.

Evaluation: What is this work's contribution to the discipline?

In the first section of the review you described the work being reviewed. You then went on to outline its strengths and flaws. Now you have to make a judgment. Is it worth reading? You may find it useful to be guided by the central question: *Would you advise people to read (or buy) the work you have reviewed?*

In evaluating the work's contribution to the discipline you should begin with the assumption that the author has something useful to say, rather than try to explain whether you agree or disagree with that contribution. What is that contribution? Has the author helped *you* to make sense of things? What has been illuminated?

Usefulness

If you have sufficient expertise in the field, **appraise** the work being reviewed as an alternative to other works that are available. It may be helpful to consult library reference material or the web to see, for example, how many other books on the same or a similar topic have been published recently. If you are reviewing a website, search for competitors and compare them.

Think about how the work you are reviewing fares against its competition.

Compare factually the work you are reviewing with its predecessors. What subjects does it treat that earlier works did not? What does it leave out? If the work has strengths or weaknesses in material that is also found elsewhere, include these comparisons. Remember to cite any additional sources you mention.

Value

A fundamental reason for writing reviews is to let readers know whether a particular book, article, or website is worth consulting. Your recommendation should be consistent with the preceding analysis of its strengths and weaknesses. For example, it would be contradictory to **criticize** a work mercilessly and then conclude by saying that it is an important contribution to the discipline and should be consulted by everyone interested in the area. It is worth restating a point made earlier: in your review *be fair, be explicit*, and *be honest*.

Written expression and presentation of the review

Assessors will consider the balance between description, analysis, and evaluation when marking a review. A description of the work is important, but it should not dominate your review. Unless there are special reasons, the description should usually be less than half of the total length of the review.

Your own writing should be clear, concise, and appropriate for your audience. Detailed advice on the assessment criteria used for written expression (which is shown in the Assessment Form for Reviews in Box 5.4) can be found in Chapter 1.

Examples of published reviews

Before you start writing a review, you will find it helpful to look to some academic journals for good examples. As a preliminary guide, Box 5.5 shows some slightly revised versions of reviews published in the *Library Journal*. You will see that, despite their brevity, the reviews illustrate many of the issues discussed in this chapter. The third review is of a website.

Box 5.5 ◉ Examples of reviews

Example 1

Meredith, M. (2005). *The fate of Africa: A history of fifty years of independence*. New York, NY: Public Affairs Perseus, 800 pp.

A scholar of Africa necessarily becomes an expert on death. In Meredith's tome, death comes in huge numbers and in many ways: through famine, ethnic strife, and racial injustice and at the hands of ruthless dictators. It came in the days of European colonialism, but in postcolonial Africa, death pervades the continent. Meredith (*Our Votes, Our Guns: Robert Mugabe and the Tragedy of Zimbabwe*) writes with sobriety, intelligence, and a deep knowledge of Africa as he describes individuals responsible for deaths unimaginable to much of the rest of the world. A well-known example is the carnage among Hutus and Tutsis in Rwanda, which claimed 800 000 lives in 100 days in 1994—more people were killed more quickly than in any other mass killing in recorded history. Much of this tragic history has been told in part elsewhere, but Meredith has compiled the text covering the entire continent. Only in the last few pages does Meredith answer the question of Africa's fate—and he thinks it's bleak. This is a valuable work for those who wish to understand Africa and its besieged peoples.

Adapted from *Library Journal* (2005), vol. 130, no. 12.

Continued

Example 2

Rossi, J. (2000). *The wild shores of Patagonia: The Valdes Peninsula and Punta Tombo*. New York, NY: Abrams, 224 pp.

Neither an experienced author nor a photographer, Rossi discovered Patagonia while on short leave from her job with the European Parliament and promptly turned her fascination into a two-year encampment to prepare this book. About half the book is devoted to a detailed discussion of southern right whales, southern elephant seals, and killer whales (including their unique beach hunting methods). Less detail is given to South American sea lions, and dusky and bottlenose dolphins. The author describes Magellanic penguins but provides limited information on 12 other kinds of birds, sometimes furnishing just one picture and caption per species. Rossi also discusses eight land animals in about 25 pages, giving short shrift to some unique creatures. Reader interest in this area may grow, since the United Nations has recently declared the Valdes Peninsula a World Heritage Site, and ecotourist visits are increasing. Although the photographs are merely average, this readable volume is recommended for members of the general public and for academic audiences interested in environmental and life sciences.

Adapted from *Library Journal* (2001), vol. 126, no. 1.

Example 3

ABC-CLIO. (2000). *World geography*. www.abc-clio.com. Date reviewed: November 15, 2000. Market@abc-clio.com, US$499 per year.

World Geography aims to be the place students go to research a number of themes and look up facts (many non-geographical) on all countries. There are three major sections providing access to the data: Home, Student, and Reference. The site includes more than 10 000 entries, including biographies, histories, maps, documents, statistics, video clips, and photographs. [At the home page] students can select a country to study; search the text by keyword or choose to catch up on world events; read a Feature Article (Why Is the Price of Gas Rising?); or answer three questions to a Where in the World? quiz. The Feature Article portion is well researched; one hopes ABC-CLIO plans to archive these essays.

In the Reference section, country selection is via a world map or drop-down menu. Text can be searched by keyword or in advanced search, where searches can be limited by type. A noteworthy inclusion here is Ask the Cybrarian, in which students can ask reference questions of ABC-CLIO.

The Student section can be used in the classroom. Students can read the syllabus, class announcements, and the word of the day and can take review tests. A nice feature here is regional overviews that cover Landforms, Climate, Vegetation and Animal Life, People, and Natural Resources and Agriculture. Other overviews cover people, news events, and organizations for the selected country.

There are also a few problems. Although country maps are clear, city maps are unfortunately excluded. Data presented are only for one year, although often more historical data are needed. News events in the Home section are listed incompletely, e.g., "Experts Investigate Possible . . ." Possible what? Forcing users to click to see if a story is worth reading is inefficient. The icon used to designate a link to a map is a small map of the United States—for all countries. Sources for news articles are not always provided, nor are clear directions on how to cite information found here. Information included is sometimes mystifying; for example, under Organizations in the United States section, corporations such as Apple, Walt Disney, AT&T, and Exxon are included. Pierre Trudeau's death is noted, but no biography is available. The Tools section offers two excellent features: the Merriam-Webster's Collegiate Dictionary and the Merriam-Webster's Collegiate Thesaurus; and ClioView, which allows students to select, sort, and view a country and topic (there are 18 from which to choose, including Literacy, Population Density, and Active Armed Forces).

The bottom line: World Geography is not exactly what its name implies. It is ABC-CLIO's successor to Exegy and is better described as a good source for beginning country studies. In that context, it is a useful resource. This site is recommended for members of the general public and for children entering secondary school.

Adapted from *Library Journal* (2001), vol. 126, no. 1.

Further Reading

Brozio-Andrews, A. (2007). *Writing bad-book reviews.* www.absolutewrite.com/specialty_writing/bad_book_reviews.htm (accessed 6 October 2009).

- This article distinguishes between a *bad-book* review and a *bad* book review, and encourages writers to try to present balanced reviews if possible.

Burdess, N. (1998). *The handbook of student skills for the social sciences and humanities* (2nd ed.). New York: Prentice Hall.

- Chapter 3 includes short sections on writing abstracts and book reviews.

Cress, J. (2013). *Writing book reviews.* http://writing.colostate.edu/guides/documents/bookreview/index.cfm (accessed 6 October 2009).

- This site contains a multi-section explanation about all aspects of writing book reviews.

Hart, C. (1998). *Doing a literature review: Releasing the social science research imagination.* London, UK: Sage.

- This is a detailed examination of literature reviews with emphasis on making arguments and organizing your ideas before writing the review.

Sova, D.B. and Teitelbaum, H. (2002). *How to write book reports* (4th ed.). Lawrenceville, NJ: Thomson-ARCO.

- This is a lower-level guide most suitable for high school students but also valid as an introduction to book reports and reviews for undergraduates.

Wilhoit, S. (2007). *A brief guide to writing from readings* (4th ed.). New York: Pearson Longman.

- This book is a helpful guide that emphasizes critical reading and development of independent ideas. Separate chapters are devoted to the skills of quoting, paraphrasing, summarizing, and critiquing.

Writing Center, University of North Carolina at Chapel Hill. (2013). *Book reviews.* http://writingcenter.unc.edu/handouts/book-reviews/ (accessed 19 November 2013).

- This site provides helpful information about writing book reviews, including examples of poor, moderate, and strong reviews for the same work.

6 Referencing and Language Matters

If the English language made any sense,
a catastrophe would be an apostrophe with fur.

—Doug Larson

A man will turn over half a library to make one book.

—Samuel Johnson

Key Topics

- What are citations and references, and why do we need them?
- Principal reference systems: author–date and note systems
- Avoiding biased language and stereotyping
- Salient notes on punctuation and spelling

One important convention and courtesy of academic communication is to cite or acknowledge the work of those people whose ideas and phrases you have borrowed. This chapter outlines the form and practice of the two reference styles used most commonly in geography and environmental sciences. It also discusses the serious matter of avoiding sexual or other biases and stereotyping in language before concluding with some notes on punctuation and spelling.

What Are Citations and References? Why Do We Need Them?

When you use information or an idea that originally appeared in someone else's work, you must acknowledge clearly where you got it (Pechenik, 2007; Turabian, 2013). In addition, you must always make the acknowledgement in a consistent and recognizable format. In your academic work you are expected to draw on evidence from, and substantiate claims with, *up-to-date*, *relevant*, and *reputable* sources. The *range of references* used for your work is also important.

In this chapter, a distinction is made between **references** and in-text **citations**. References provide the full **bibliographic details** for a source and are collected in a **reference list** at the end of a chapter or complete work. In-text citations are inserted into the text and contain just enough information to provide an unambiguous link to the full reference.

A discussion about the use of sources in writing is found in Chapter 2. To summarize briefly, you must cite all references in order to

- acknowledge previous work conducted by other scholars;
- allow the reader to verify your source materials; and
- provide information so that the reader can consult your sources independently.

You *must* clearly acknowledge your references when you *quote*, *paraphrase*, or *summarize* information, ideas, text, data, figures, tables, or any other material that originally appeared in someone else's work. Acknowledgements may be given to sources such as books, journal articles, newspaper stories, maps, films, photographs, reports, websites, or personal communications, such as letters or conversations. References must contain enough bibliographic details for your reader to be able to find your source easily. *Easily* in this instance means that the source is identified unambiguously, although actually acquiring that reference may not be easy for another person (e.g., you may have consulted old books or articles in obscure journals). See Box 6.1 for a discussion of the use of software programs to collect, store, and organize bibliographic details of your references.

Citing sources fully and correctly is an important academic skill. Take the time to learn how it is done.

Principal Reference Systems: Author–Date and Note Systems

There are three principal reference systems: the **author–date system** (sometimes called the **Harvard**, in-text, or scientific system); the **note system** (sometimes called the endnote or footnote system); and the *Vancouver* system. Of the three systems, the author–date system (and variants of it) is the most widely used in geography and the environmental sciences. The sample paper in the Appendix was prepared using the author–date system. The note system, although less common, is employed by some publications in geography (particularly in the humanities). For this reason, the note system

Box 6.1 ◉ Bibliographic-management software programs

Software programs are available for storing and managing bibliographic details for reference sources. There are a number of good reasons for using a bibliographic-management program:

- *Electronic storage.* No longer do you have to copy out bibliographic details from pieces of paper that may get scattered among your belongings, or lost—a database is a single place where you can store all your references.
- *Records stored for repeat use.* For undergraduate course work, you may collect references for one-time use, such as a single essay or report. However, you may also find that you use certain references a number of times throughout your undergraduate program or if you go on to graduate work or an academic career. A program allows you to accumulate the details of references you have collected in a personal, searchable database.
- *Direct transfer from research databases.* Although you have the option of entering bibliographic details manually, programs can import details directly from online research databases. This will ensure that your records are accurate and not prone to mistakes that you might make when typing. But since not all types of references are summarized in databases that permit direct transfer, some manual entry work is likely to be necessary.
- *Efficiency during research.* Previously this book mentioned that you should keep track of the bibliographic details of your references as you proceed with your research. Direct transfer from an online database to the program is usually as easy as making a few clicks of a mouse. This is much more efficient than writing out or typing details, allowing you to get on with your research.
- *Accurate reference lists.* In addition to storing bibliographic details, these programs can output reference details you select in a wide variety of reference formats. If you pick a suitable output format and the information is complete and correct, creating a reference list can be much faster and easier than doing it manually.

A large number of bibliographic-management programs are available, many of them free (search the web for "bibliographic management software"). Standalone and web-based versions are available, the latter being particularly useful because they permit access from any computer with Internet access. Three noteworthy programs are EndNote, ProCite, and RefWorks. None of these are free; while individuals can purchase the software, your institution may provide free access through a group or site licence. Ask at your institution's library to see if a licence is held and if you are eligible to register for an account. These particular programs are mentioned because they are commonly integrated directly into online databases, thus permitting easy, direct transfers of bibliographic information to your personal database.

> *Most writing in geography and environmental sciences employs the author–date reference system.*

is also discussed in the pages that follow. Because the Vancouver system is confined largely to medicine and other health-related disciplines, it is not discussed here. If you need to use this system, consult the International Committee of Medical Journal Editors website (ICMJE, 2013) for advice.

Variants of reference systems: Reference styles

Within the author–date and note systems, there are numerous variants that have differences (often very subtle differences) in how references are cited and listed (Library of Congress, undated). These are called **reference styles**. One commonly used style is APA (American Psychological Association). For the author–date system, reference style also includes in-text citations, as discussed below. The following are some examples of variations in reference style:

- The title of a journal article may or may not be in quotation marks.
- Journal volume and issue numbers may or may not be printed in boldface or italics, respectively; issue numbers may or may not be included.
- Titles of works may or may not be italicized.
- The way in which titles of works are capitalized can vary.
- The number of pages in books may or may not be listed.
- The punctuation (or lack of punctuation) used to separate bibliographic elements may differ.
- For the second and subsequent authors in a list, the initials of the first name may either precede or follow the surname.

These variations in referencing style can be confusing for a writer who is expected to prepare a reference list. What is the correct style? The answer is that there is no single correct style. Here are some tips to help you approach the matter of reference style:

- *Remember the purpose of references*. This is the most important consideration. You need to inform readers of the key bibliographic details of your sources so that they are able to identify accurately (and possibly seek to acquire) those sources from the information you provide. The content of references must be your first concern; style is secondary.
- *Follow style guidelines, if provided.* Paying attention to reference style should come after you have ensured that the necessary bibliographic

details have been included and are accurate. If the assignment description includes specific instructions about reference style or a list of examples, there is a single correct style—for that situation. Follow any supplied instructions carefully even if the details are different from styles you have used before. For example, you may be told simply to "use APA style" (a popular variant of the author–date system), or another standard reference style, and be expected to seek resources about that style on your own. Or you may be given a set of instructions (usually with examples) describing a specific style of formatting which you are expected to use to format the bibliographic details of your references. For example, academic journals, such as the *Canadian Journal of Earth Sciences* (2013), usually provide a set of author guidelines that defines the required reference format that must be followed for an article to be accepted for publication.

- *Choose a reference style to use as your personal default.* Find a valid author–date reference style and a note reference style (as mentioned, there is no lack of standards that you can use as your example—including the style used in this book), and become familiar and comfortable with using them. Then, if your assignment description just says, "References must be cited and listed properly," you can use your chosen system. You should expect to be marked down for not including a reference list, not presenting your list neatly and consistently (see the next point), or not following the reference-style instructions if they are provided. However, if you did submit a reference list that contained the important bibliographic information and was styled consistently, it would be highly unfair of a professor to complain that you did not use a specific system or style if you were not made aware of that requirement.
- *Ensure the reference list style is consistent.* Use a consistent style throughout the reference list. Paying attention to details and consistency in the reference list is one sign of quality work and sends a positive message to your reader. If the style you are using calls for titles of journals and books to be italicized, make sure this is done with all such titles and that other elements of references are not italicized randomly. After you have prepared the reference list, go through it again looking specifically for inconsistencies in type style, punctuation, order of elements, and so on. A common problem in the digital age is that writers electronically cut and paste references that others have prepared with a range of styles, and then fail to revise a list of references into a consistent style.
- *If you are unsure how to cite a source, make an effort to learn how to do so.* The majority of your references may be to common types of sources, such as

books, chapters in books, journal articles, or websites, and through repetition you will become familiar with formatting those types of references. However, you may need to cite other, less common, types of sources. If you are unfamiliar with the correct style for a particular type of source, either (a) search for examples in the system you are using or (b) make your own decision about how the reference should be cited so that it is consistent with the rest of your list and gives the reader the important bibliographic details. In the following sections of this chapter, examples are given of a number of less common sources for writers in geography and the environmental sciences.

The author–date (Harvard) system

The author–date system comprises two essential components: brief *in-text citations* throughout your work and a comprehensive *list of references* at the end of the work.[1] The in-text citation gives the surname(s) of the *author(s)* and the *date* of publication. If the reference is to specific information (including quoted passages, specific points, or facts), the *pages* where the information or quotation can be found are included. Similarly, if the reference is to a figure or table the figure or table number is included. If the reference to the source is less specific, the pages may be omitted. Further information about the source is not required in the text because the citation tells the reader that the full bibliographic details are provided in the list of references. This information differs according to the type of source, as discussed more fully below.

In-text citations

The in-text citation presents only a brief summary of bibliographic details. Take particular note of the ways in which the various citations are punctuated in the examples that follow, but keep in mind that the discussion about variations in reference-list style also applies to in-text citations.

See Chapter 2 for a discussion of how to incorporate citations into the text. The sample paper in the Appendix applies information presented in that discussion. Box 6.2 lists some of the most common formatting errors that are made when using the in-text citation system.

---

[1] Footnotes may be used within the author–date system for an aside or for information supplementary to the main text. Indeed, this is an example of such a footnote. Details on the use of footnotes are provided in the section "Substantive notes" later in this chapter.

Box 6.2 ◉ Common in-text citation formatting errors

The models for formatting in-text citations shown in this part of the chapter should be easy to follow. To help further with learning this skill, here is a list of six common errors that are made.

1. Inserting the citation outside of the sentence in which it belongs.

   Incorrect: The mean thickness of Arctic Ocean ice cover is decreasing. (Bain, 2012)
   Correct: The mean thickness of Arctic Ocean ice cover is decreasing (Bain, 2012).

   Note where the period is placed in relation to the citation.

2. Neglecting to include second or subsequent authors. In the following example, the work cited has two authors.

   Incorrect: Discharge patterns on the Rapid River were studied by Jones (2006).
   Correct: Discharge patterns on the Rapid River were studied by Jones and Roy (2006).

   Similarly, if a 2006 publication was written by Chen, Hibbert, Perez, and Woods, the in-text citation should be to Chen et al. (2006), not just Chen (2006).

3. Omitting source page number(s) for quoted material.

   Incorrect: Armstrong (1997) stated that "Potential errors in estimating the size of bat colonies have been neglected in previous studies."
   Correct: Armstrong (1997, p. 763) stated that "Potential errors in estimating the size of bat colonies have been neglected in previous studies."

   If a figure or table is based on a source, the figure or table number in the source publication (instead of the page number) should be indicated in the citation you include in the caption.

4. Including source page number(s) for non-specific material that is not quoted.

   Incorrect: Armstrong's (1997, p. 763) work was the first study of bat populations in this part of Ontario.
   Correct: Armstrong's (1997) work was the first study of bat populations in this part of Ontario.

   A page number can be given for non-quoted material if you want to point the reader to specific information. In this example, the reference

Continued

is essentially to the entire Armstrong publication, so no page number would be given.

5. Omitting source's year of publication.

   Incorrect: Greater Vancouver's population grew faster than Greater Montreal's during the study period (Baker). Comparative rates of growth were . . .

   Correct: Greater Vancouver's population grew faster than Greater Montreal's during the study period (Baker, 2001). Comparative rates of growth were . . .

6. Omitting a citation completely when presenting illustrative material reproduced or modified from a source.

   Just because an illustration is not written text does not mean that a citation can be omitted. Indicate either "Reproduced from . . ." or "Adapted from . . ." in your caption, and include the figure or table number in the source publication in the citation.

- Reference is to a work in general:

  Depletion of groundwater aquifers is a serious problem in many parts of the world (Roberts, 2009).

Note that some reference styles place a comma between the author and publication date, but others do not. This is an example where it is useful to have accepted standards to follow in a discipline, but it is difficult to argue technically for or against many specific aspects of particular styles. In the absence of specific guidelines, the important point is that the same style should be used consistently throughout a work.

Use one system of referencing consistently throughout a piece of work.

- Reference is to quoted material:

  "Philosophy in physical geography is done as much by boots and compass as by mental activity" (Inkpen, 2005, p. 1).

Include the page number on which the quoted material appears in the source.

When referring to or using specific material from the source in your work without quoting it directly, supplement the citation with the location of the material. The following three examples show how to do this.

- Reference is to *specific* information on *one page* or on *more than one page*:

  As Bloggs (2006, p. 50) has made so clear, a significant challenge confronting geography is . . .

  Environmental management is an area of growing employment opportunity that has attracted many new graduates in the past decade (Jones, 2005, pp. 506–507).

  If the citation is to material with a wide page range or spread over many pages, it is probably a reference to a more general idea or message, in which case the page numbers can be omitted. Furthermore, many common citation styles do not include page references for any material that is not quoted. APA is one well-known style that does dictate the inclusion of page references for most citations to specific, paraphrased material.

- Reference is to a *figure* or *table*:

  Howard (1998, Fig. 3) documented the relationship between . . .

  This includes any figure or table that you mention, reproduce, or modify; or contains data that you have used in the creation of your own figure or table.

- Reference is to a *number of sources*:

  Several authors (Henare, 2004; Nguyen, 2005; Brown, 2006) agree . . .

  Note that each reference is followed by a semicolon. Here is a common alternative format with less punctuation:

  Several authors (Henare 2004, Nguyen 2005, Brown 2006) agree . . .

  Multiple citations may be listed in chronological order, reverse chronological order, or alphabetically by author's name.

- Reference is to a single work written by *two people*:

  A recent study (Chan and D'Ettorre, 2002) has shown . . .

  It has been made clear (Brown and McDonald, 2006) that . . .

  Smith and Black (2002) have shown that . . .

  Note that the names are linked within parentheses by the word *and*. Some styles use an ampersand (&) to link names shown within parentheses but use an *and* if the names are incorporated within the text.

- If the reference is to a single work written by *more than three people*:

  Harvey et al. (2003) argue that . . .

  The abbreviation **et al.** is short for *et alii*, meaning "and others." The term may be written in italics although it need not be. Its use varies: some styles use *et al.* when there are *more than three* authors, not two.

- Reference is to an anonymously written work:

  This is apparently not the case in Thailand (*Far Eastern Economic Review*, 2003, p. 12).

  An editorial in *The Globe and Mail* (1998) offers fine testament to this view.

  The expressions *Anonymous* and *Anon.* should not be used. Instead, a publication's title is given in place of an author's name.

An in-text citation is a short summary of the bibliographic details. Full details of each reference are given in a separate list of references at the end of your work.

- Reference is to a *map* without an author named:

  Low levels of precipitation are evident through much of southeastern Alberta and southwestern Saskatchewan (*Atlas of Canada*, 2002).

- Reference is to work written by a *committee* or an *organization*:

  Agriculture Canada (2004, p. 41) suggests that soil degradation is of major concern to the agricultural community in Saskatchewan.

  Natural disasters may present significant difficulties for residents of British Columbia (Geological Survey of Canada, 2007).

- Reference is to one author referred to in the writings of another:

  Motor vehicles are a major cause of noise pollution in urban areas (Hassan, 2003, cited in Yeo, 2005, p. 219).

  Give the page number on which the citation appears. Avoid such references unless it is impossible to trace the original source. You are expected to find the original source yourself (in this case, Hassan, 2003) to ensure that the information has not been misinterpreted or misquoted by the intermediate author (Yeo, 2005). However, you should include both sources in the reference list so your reader can search for either or both.

- Reference is to information gained by means of *personal communication*:

  Aspects of the theory await investigation (Wotjusiak, 2006, pers. comm., May 23).

  References to personal communications should be incorporated more fluidly into the text than in the example above and, where appropriate, should include some indication of the quoted person's claim to authority or expertise in the relevant field:

  In an interview conducted on 12 January 2006, Dr Melody Jones, director of City Services, revealed that . . .

  Personal communications are not usually included in the list of references, but if you have cited a number of personal communications, it may be useful to provide a separate list that gives the reader some indication of the credibility of the people cited (that is, some indication of their authority in relation to your work).

- Reference is to electronic information:
  The style of reference is the same as for individual, group, organizational, and committee authors, as outlined above. The in-text citation should give the author's name (which may be an institution) and the year in which the reference was created or last amended (for websites this is usually noted toward the end of a page as "Page last updated on . . .").

  The National AIDS Information Clearinghouse (2002) guidelines give clear advice on . . .

  If there is no indication of a production date, you should record this as "undated."

  Genetic engineering seems likely to have a profound influence on global crop production figures (Witherspoon, undated).

  The in-text citation for an online source should *not* include the URL of the site or page. This information is included only in the reference list.

Reference list

This *alphabetically* ordered list (by surname of author) provides the complete bibliographic details of all sources actually referred to in the text. By convention, however, it does not include those sources you consulted but have not cited (a full list of *all* references consulted is known as a *bibliography*).

The following examples of references, formatted with a modified version of

APA style, cover a range of commonly encountered sources and may be useful when you prepare your own reference lists. (Anglia Ruskin University [undated] and the Library of Congress [undated] have published detailed guides to the author–date system, with instructions and examples for virtually any type of source that you can imagine.) These are examples of *non-electronic sources*, such as books or journal articles published in hard-copy form. References for electronic sources are usually versions of these styles suitably adapted to include the electronic access details. Examples of references to electronic sources are found below.

Look carefully at the examples and distinguish between the various kinds of work and how each is organized and punctuated. Note the following:

- The second and subsequent lines of each reference are indented. (To keep the indents in the right place, use the program's "hanging" or "hanging-indent" paragraph format instead of inserting tabs manually.)
- Reference lists are single-spaced but with a blank line after each entry.
- Titles of books, chapters, and articles are given minimal capitalization, but journal titles are more fully capitalized.
- Titles of single published works (e.g., books, journals, and maps) are italicized. Parts of a publication (e.g., a journal or newspaper article or a chapter in a book) are *not*.

Complete book

Blunt, A., and Wills, J. (2000). *Dissident geographies: An introduction to radical ideas and practice*. Edinburgh Gate, UK: Pearson Education.

Note that all authors' surnames precede their initials in a list of references. When multiple authors are listed, a comma is placed between the final *and*, and the preceding initial. All authors should be named in the reference list, even if you have used *et al.* for works with more than three authors in your in-text citation.

Book, edition other than first

Ritter, D.F., Kochel, R.C., and Miller, J.R. (2002). *Process geomorphology* (4th ed.). New York: McGraw-Hill.

Edited volume

Pickles, J. (ed.). (1995). *Ground truth: The social implications of geographic information systems*. New York: Guilford Press.

Most references to edited collections will, in fact, be to specific chapters rather than to the book as a whole. For that reason, it is more usual to see those references presented in the form of a "chapter in an edited volume" (shown below).

Chapter in an edited volume

Clague, J.J. (1991). Natural hazards. In H. Gabrielse and C.J. Yorath (eds.), *Geology of the Cordilleran orogen in Canada.* Ottawa: Canada Communications Group (pp. 803–815).

Note that the editor's surname and initials are not inverted.

Article in a journal

Journals are **periodicals** typically intended for academic use. Journals commonly give greater emphasis to volume and issue number than to their week or month of publication (which is stressed with popular magazines).

Brottem, L., and Unruh, J. (2009). Territorial tensions: Rainforest conservation, postconflict recovery, and land tenure in Liberia. *Annals of the American Association of Geographers, 99*(5), 995–1002.

Article in a magazine

Magazines follow a similar style to journals, but the volume and issue number are replaced by details of the publication date.

Brown, A. (2006, January 16). Muddied waters. *Gourmet Traveller Wine*, pp. 23–26.

If the article is continued later in the magazine, give all page numbers.

Loi, V. (2006, January 16). Tales from Perth. *Australian Traveller*, pp. 23–28, 68.

Paper in conference proceedings

Bakke, B., and Sneed, W. (2008). When methanol addition is cheaper than free wastewater for denitrification. In Water Environment Foundation, *Proceedings of the Water Environment Federation, WEFTEC 2008* (pp. 7602–7620). Chicago, IL.

Paddock, L. (2005). Strategies and design principles for compliance and enforcement. In International Network for Environmental Compliance and Enforcement, *Proceedings of the Seventh International Conference on Environmental Compliance and Enforcement, Vol. 1* (pp. 67–72). Marrakesh, Morocco.

Publication with no author identified

Statistics Canada. (2009). *How to Cite Statistics Canada Products*, Cat. no. 12-591-X. Ottawa: Statistics Canada.

If the author is an organization, the name can be abbreviated when it appears as the publisher if there is a common abbreviation (e.g., Australian

Bureau of Statistics is well known as ABS; United States Geological Survey is also known as USGS).

Thesis or dissertation

Van Loon, T. (1997). *Regeneration in logging gaps in the moist forest of southern Cameroon*. Master's thesis, Department of Forestry, Wageningen Agricultural University, Wageningen, The Netherlands.

Witterseh, T. (2001). *Environmental perception, SBS symptoms and the performance of office work under combined exposures to temperature, noise and air pollution*. Doctoral dissertation, Technical University of Denmark, Lyngby, Denmark.

Be sure your references are styled correctly. Many markers (and editors) are very particular about this.

Unpublished document

Curtis, K., and Wagner, M.C. (2008). *Cougar Creek sediment control study* (Draft). Unpublished report prepared for the Town of Canmore by AMEC Earth and Environmental, Calgary.

Water Resources Branch (WRB). (1997). *Manitoba Water Resources Branch daily water levels and forecasts, 17 April–7 May 1997*. Unpublished manuscript, Manitoba Conservation, Water Resources Branch. Winnipeg.

Unpublished materials come in a variety of forms, such as letters and papers presented at conferences. Provide, in a concise way consistent with the style of other references, sufficient details for your reader to be able to find the material cited.

Newspaper article

Cornacchia, C. (2006, September 29). Global warming being linked to blue-green algae. *Winnipeg Free Press*, p. A1.

If no author is identified clearly, use the newspaper's name as the "author" but do not repeat it after the article title.

Media release

Patterson, B. (2009, September 29). *Saint Mary's installs Atlantic Canada's first living wall*. Media release, Saint Mary's University, Halifax.

Video, television broadcast, film

Nova (2009, December 8). *Volcano above the Clouds 2009* [television broadcast]. PBS.

An Inconvenient Truth (film) (2006). Paramount Vantage, Hollywood, CA.

CD-ROM

Material on a CD-ROM is dealt with in much the same way as comparable hard-copy publications, such as books or journals, except that a CD-ROM statement is inserted after the CD's title.

Pawson, E. (1999). Remaking places. In *Explorations in Human Geography: Encountering Place* (CD-ROM), eds. R. Le Heron, L. Murphy, P. Forer, and M. Goldstone. Auckland, New Zealand: Oxford University Press.

Software program

Author's surname, Initial(s) (or corporate author), Date, *Title of program*, Version, Publisher, City [if available].

Adobe Systems (2008). *Acrobat*, v. 8.0, Adobe Systems, San Jose, CA.

Electronic Arts (2000). *SimCity 3000 Unlimited*, Electronic Arts.

Map

Mapping organization and date, *Map title*, Edition number (if appropriate), Scale, Series, Publisher, Place.

Centre for Topographic Information. (2000). *Sherbrooke*, 7th ed., 1:50,000, National Topographic Series. Natural Resources Canada, Ottawa.

Aerial photograph

In most cases, an aerial photograph will not have a formal title but can be identified by flight index and frame numbers. The general style for a reference to an aerial photograph is as follows:

Custodian and Year of Photography, *Title* (may be described by photograph's flight path and index number), Scale, (medium of the source), Publisher, Date of photography.

NAPL (1981). *A25767-66*, 1:12,000 (black-and-white aerial photograph). Ottawa: Natural Resources Canada, National Air Photo Library, June 10, 1981.

Satellite or airborne imagery

A reference structure similar to the one used for an aerial photograph can be used for satellite or airborne imagery. Custodian, sensor type, date of image collection, and medium of the source are key elements to incorporate in these references. If other descriptive information (e.g., satellite path and row) helps characterize the image and is available, it can also be included.

GeoGratis (2003). *Landsat 5 TM image of Winnipeg, Manitoba* (satellite image), Landsat 5 TM Data over Major Canadian Cities. Natural Resources Canada.

In this example the image is held and distributed by GeoGratis, which therefore is listed as the "author."

References to Internet sources (author–date system)

There has been a proliferation of materials available in electronic form, and these sources present new challenges for referencing. Many older publications describing reference styles do not include references to Internet and other digital resources. (For example, before the early 1990s the Internet did not have the ubiquitous presence that it does today, so it was simply not a source of reference materials.) However, the range of information resources now includes new categories of source materials; as a result, information on this topic is no longer scarce, either in more recent publications (Turabian, 2013) or on the web (e.g., Library of Congress, undated; Anglia Ruskin University, undated).

For most of the types of sources listed above, the reference format can be modified to include details of the electronic format as well as the date the material was accessed. Any elements of the reference format that are specific to hard-copy materials are omitted. Some examples will be given for the most common types of electronic sources you will encounter. For other types of sources acquired in electronic form, use your own judgment to adapt the reference format appropriately, or search for examples.

Remember that including essential bibliographic information for the reader, following consistent formatting, and using logical organization are the paramount aspects of creating a reference list. Beyond that, following a particular style becomes important when you have been given specific instructions to follow.

PDF documents

Some PDF documents are simply electronic versions of materials available in hard-copy form, whereas others are available exclusively in electronic form. If you have obtained a PDF online, include relevant location or access details.

Department of Health (2008). *Health Inequalities: Progress and Next Steps* (PDF online). London, UK. http://www.dh.gov.uk/en/Publicationsandstatistics/Publications/PublicationsPolicyAndGuidance/DH_085307 (accessed November 27, 2009).

Najam, A., Runnalls, D., and Hallam, M. (2007). *Environment and Globalization: Five Propositions* (PDF online), International Institute for Sustainable Development. www.iisd.org/pdf/2007/trade_environment_globalization.pdf (accessed November 27, 2009).

Websites and web pages

If you have decided that a selection of material found online (for example, on a website) is a valuable source for your work and can be considered credible (see Chapter 2 for discussion of this topic), you must reference it just like any other source. There are many examples of valid web resources, such as databases, descriptive web pages, and images, that do not fit neatly into any source category discussed so far. However, first see if the material can be assigned to one of the types previously discussed. If so, modify the reference format to reflect the online location. Otherwise use these examples as models for references to general online sources:

> ***Electronic sources present some referencing challenges. Be sure to provide sufficient information to allow someone else to find the source.***

Environment Canada. (2009). *Canadian Climate Normals or Averages 1971–2000.* www.climate.weatheroffice.ec.gc.ca/climate_normals/index_e.html (accessed November 27, 2009).

Time and Date AS. (2009). *The World Clock—Time Zones.* www.timeanddate.com/worldclock (accessed November 27, 2009).

The Cook Islands. (Undated). www.ck (accessed November 27, 2009).

Electronic books

An increasing number of books are now available online, and a book may be published exclusively in electronic form. The reference format for a hard-copy book can be adapted to include the electronic location or access details:

Bowers, C.A. (2006). *Transforming environmental education: Making the cultural and environmental commons the focus of educational reform.* n.p.: Ecojustice Press. http://hdl.handle.net/10535/15 (accessed November 27, 2009).

Translation Bureau. (1997). *The Canadian Style: A guide to editing and writing.* Toronto: Dundurn Press. http://site.ebrary.com.library.smu.ca:2048/lib/smucanada/docDetail.action?docID=10221229 (accessed November 27, 2009).

Electronic journal articles

There has been a proliferation of articles made available on the Internet through electronic journal databases (usually with restricted access granted by a library or through a personal subscription). The format shown above for journal articles can be used for articles accessed in print or electronic versions that have equivalent bibliographic information. That is, the electronic version accessed on the web may be identical to the print version, except that it is in a digital form (commonly a PDF document).

Some reference styles (and professors) require you to include the Internet address or electronic database for an article acquired online or to include the generic note "online" to indicate that is how you accessed it. Many articles available electronically are now being identified with a unique "**DOI**" (Digital Object Identifier), which may be also included in the reference.

Format information as consistently as possible with other references in the list.

Journal articles published exclusively online may not have exactly the same information as articles that come as print versions. For example, a consecutive page range may not be considered necessary in an online journal. Instead, an article may be shown as having a certain number of pages. To help a reader find an online article, include some information about its Internet location or the electronic database.

Caldwell, D.R. (2005). Unlocking the mysteries of the bounding box. *Coordinates*, Series A, no. 2. http://purl.oclc.org/coordinates/a2.htm (accessed November 27, 2009).

Morrison, A. (2007). The geospatial content of public transport websites for 60 localities worldwide. *Journal of Maps*, vol. 2007, 1–19. doi:10.4113/jom.2007.76 (accessed November 27, 2009).

Email, discussion lists, and newsgroups

Privacy and dissemination of personal details are concerns in this area. It is good practice to assume that all messages sent through Internet networks enter the public domain, even if they were intended to be personal communications. From this perspective you can argue that it is valid to include someone's personal details in your reference. Anglia Ruskin University (undated) suggests contacting the person for permission. Messages can be presented in the following format:

Author's surname, Initial(s). author's email address, Message year, "subject line from message" (medium of the source), Description of message or statement of list name (if appropriate), List or recipient address, Date of message.

Email addresses (and Internet addresses) need not be underlined, but you will find that many word-processing packages do this automatically.

Thomson, K. k.thomson@massive.com.au. (2005). Historical Geographies of South Australia (email). IAG-list, iag-list@ssn.flinders.edu.au. April 9.

Brown, W. william.brown@hotmail.com. (2006). Job opening here (email). Personal email to Melanie Smith, m.smith@msn.com. June 3.

Multiple entries by single author

If you have cited two or more works written by the same author, you should present them in the reference list in chronological order by date of publication. If they were written in the same year, add lower case letters to the year of publication in both the reference list *and* the text to distinguish one publication from another (e.g., 1987a, 1987b). List the same-year publications alphabetically according to the initial letters of significant words in the reference's title, and assign letters accordingly. The following example illustrates both same-year publications and same-author publications.

Slaymaker, O. (2000). Assessing the geomorphic impacts of forestry in British Columbia. *Ambio, 29*(7), 381–387.

——— (2001a). The role of remote sensing in geomorphology and terrain analysis in the Canadian Cordillera. *International Journal of Applied Earth Observation and Geoinformation, 2001*(1), 11–17.

——— (2001b). Why so much concern about climate change and so little attention to land use change? *Canadian Geographer, 45*(1), 71–78.

As the example above illustrates, where more than one reference by an author is included, the author's name can be replaced by a three-em rule (———) in subsequent references.

Multiple entries with the same lead author

You may have more than one reference in which the same person was the lead (first) author with one or more different co-authors. Place any sole-author references first in the order described above. Then list multiple-author references in alphabetical order by the second author's surname, and so on. In cases where there is more than one source with an identical list of multiple authors, use the chronological rules for ordering the references as described in the preceding example.

The note system

This system of referencing provides your reader with note or numerical references to a series of footnotes or a list of endnotes. As described below, these notes list the bibliographic details of each reference cited. For that reason, you (and your professor) may consider it unnecessary to include a consolidated list of references at the end of your work. Whether to include a bibliography is a matter of choice in the note system. However, it is worth noting that for the reader of your work, a bibliography can certainly be helpful for locating specific references you have used.

In-text references, footnotes, and endnotes

At each point in the text where you have drawn upon someone else's work, or immediately following a direct quotation, place a superscript numeral (e.g., [3]). This refers the reader to full reference details provided either as a **footnote** (at the bottom of the page) or as an **endnote** (at the end of the document or chapter). If you refer to the same source seven times, there will be seven separate note identifiers in the text referring to that source. You must also provide seven endnotes or footnotes. This practice is sometimes simplified by giving full bibliographic details only in the first endnote or footnote and abbreviated details in the subsequent notes (Translation Bureau, 1997, p. 174).

However, the shortened notes should leave your reader in no doubt as to the precise identity of the reference. (For example, if there are two books by the same author, you will need to give enough details for the reader to know which one you mean.) In some disciplines, Latin abbreviations such as **ibid**. (from *ibidem*, meaning "in the same place"), **op. cit.** (from *opere citato*, meaning "in the work cited"), and **loc. cit.** (from *loco citato*, meaning "in the place cited") are used as part of that abbreviation process, although this practice is far less common today than it was in the past.

Note the following conventions:

- The first endnoted or footnoted reference to a work must give your readers all the bibliographic information they might need to find the work.
- Article and chapter titles are given minimal capitalization but book and journal titles are capitalized.
- Titles of single published works (such as a book, journal, or map) are italicized. Do *not* italicize parts of a single publication, such as a journal article, newspaper article, or chapter in a book.
- Article and chapter titles appear in quotation marks.
- The page number(s) for the material being cited should be included in each note. The bibliography will record the full page range of any article or chapter drawn from a larger work (such as a book or journal).

Complete book

1. A. Blunt and J. Wills, *Dissident geographies: An introduction to radical ideas and practice* (Harlow: Pearson Education, 2000), p. 102.

Book, edition other than first

2. D.F. Ritter, R.C. Kochel, and J.R. Miller, *Process geomorphology*, 4th edn. (New York: McGraw-Hill, 2002).

Edited book

3. J. Pickles (ed.), *Ground truth: The social implications of geographic information systems* (New York: Guilford Press, 1995).

Most references to edited collections will, in fact, be to specific chapters within the volume rather than to the volume as a whole. For that reason, it is more usual to see references presented in the form of a "chapter in an edited volume" (shown below).

Chapter in an edited volume

4. J.J. Clague, Natural hazards, in *Geology of the Cordilleran orogen in Canada*, eds. H. Gabrielse and C.J. Yorath (Ottawa: Canada Communications Group, 1991), p. 807.

Article in journal

5. L. Brottem and J. Unruh, Territorial Tensions: Rainforest Conservation, Postconflict Recovery, and Land Tenure in Liberia, *Annals of the American Association of Geographers*, vol. 99, no. 5, 2009, p. 998.

Article in a magazine

6. A. Brown, Muddied waters, *Gourmet Traveller Wine*, January 16, 2005, pp. 23–26.
7. V. Loi, Tales from Perth, *Australian Traveller*, January 16, 2006, pp. 23–28, 68.

The second example shows what to do if the article runs on to later pages in the magazine.

As you are preparing your work, keep full details of all the references you consult. This will make it easier to prepare a reference list.

Publication with no author identified

8. Statistics Canada, *How to Cite Statistics Canada Products*, Cat. no. 12-591-X, Statistics Canada, Ottawa, 2009.

Paper in conference proceedings

9. B. Bakke and W. Sneed, When methanol addition is cheaper than free wastewater for denitrification. In Water Environment Foundation, *Proceedings of the Water Environment Federation, WEFTEC 2008*, Chicago, Illinois, 2008, pp. 7602–7620.

10. L. Paddock, Strategies and design principles for compliance and enforcement. In International Network for Environmental Compliance and Enforcement, *Proceedings of the Seventh International Conference on Environmental Compliance and Enforcement, Vol. 1*, Marrakech, Morocco, 2005, pp. 67–72.

Thesis or dissertation

11. T. Van Loon, *Regeneration in logging gaps in the moist forest of southern Cameroon*, MSc thesis, Department of Forestry, Wageningen Agricultural University, Wageningen, The Netherlands, 1997, p. 42.
12. T. Witterseh, *Environmental perception, SBS symptoms and the performance of office work under combined exposures to temperature, noise and air pollution*, PhD thesis, Technical University of Denmark, Lyngby, Denmark, 2001, p. 104.

Unpublished document

13. K. Curtis and M.C. Wagner, *Cougar Creek Sediment Control Study* (Draft), unpublished report prepared for the Town of Canmore by AMEC Earth and Environmental, Calgary, 2008, p. 16.
14. WRB (Water Resources Branch), *Manitoba Water Resources Branch daily water levels and forecasts, 17 April–7 May, 1997*, unpublished manuscript, Manitoba Conservation, Water Resources Branch, Winnipeg, 1997, p. 7.

Newspaper article

15. C. Cornacchia, Global warming being linked to blue-green algae, *Winnipeg Free Press*, Winnipeg, Manitoba, September 29, 2006, p. A1.

If no author is named clearly, insert the newspaper's name as the "author" but do not repeat it after the article title.

Media release

Aside from the location of the author's initials and year of publication, media releases are presented in the same way as under the author–date system:

16. B. Patterson, *Saint Mary's Installs Atlantic Canada's First Living Wall*, media release, Saint Mary's University, Halifax, Nova Scotia, September 29, 2009.

DVD, television broadcast, movie

17. *Volcano above the Clouds* (television broadcast) 2009, Nova, PBS, December 8, 2009.
18. *An Inconvenient Truth* (movie), Paramount Vantage, Hollywood, California, 2006.

CD-ROM

Material on a CD-ROM is dealt with in much the same way as that for comparable hard-copy publications, such as books or journals, except that a CD-ROM statement is inserted after the CD's title.

19. Pawson, Remaking places, in *Explorations in Human Geography: Encountering Place* (CD-ROM), eds. R. Le Heron, L. Murphy, P. Forer, and M. Goldstone, Oxford University Press, Auckland, New Zealand, 1999.

Software program

Try formatting these as follows:

Author's initial(s) Surname, (or corporate author), Date, *Title of program*, Version, Publisher, Place of publication [if available].

20. Adobe Systems, *Acrobat*, v. 8.0, Adobe Systems, San Jose, California, 2008.
21. Electronic Arts, *SimCity 3000 Unlimited*, Electronic Arts, 2000.

Map

References to maps are usually presented in the following format:

Mapping organization and Date, *Map title*, Edition number (if appropriate), Scale, Series, Place, Publisher.

22. Centre for Topographic Information, *Sherbrooke*, 7th edn., 1:50,000, National Topographic Series (Ottawa: Natural Resources Canada, 2000).

Aerial photograph

In most cases, an aerial photograph will not have a formal title but it can be identified with flight index and frame numbers. Here is the general format for a reference:

Custodian and Year of Photography, Title (may be described by photograph's flight path and index number), Scale, (medium of the source), Publisher, Date of photography.

23. National Air Photo Library, *A25767–66*, 1:12,000, (black-and-white aerial photograph), Natural Resources Canada, June 10, 1981.

Satellite or airborne imagery

A reference structure similar to the one used for an aerial photograph can be used for satellite or airborne imagery. Custodian, sensor type, date of image collection, and medium of the source are key elements to incorporate in these references. If other descriptive information (such as satellite path and row) that helps characterize the image is available, it can also be included.

24. GeoGratis 2003, *Landsat 5 TM image of Winnipeg, Manitoba* (satellite image), Landsat 5 TM Data over Major Canadian Cities, Natural Resources Canada, 2003.

Since the image is held and distributed by GeoGratis, it is listed as the "author."

References to Internet sources (note system)

The section about the author–date system of referencing in this chapter included a discussion of reference styles adapted to fit materials in electronic form acquired on the Internet. Examples of references to Internet sources in the note system are given below without repeating most of the previous discussion.

PDF documents

25. Department of Health, *Health Inequalities: Progress and Next Steps* (PDF online), London, UK, 2008, http://www.dh.gov.uk/en/Publicationsandstatistics/Publications/PublicationsPolicyAndGuidance/DH_085307 (accessed November 27, 2009).
26. A. Najam, D., Runnalls and M. Hallam, *Environment and Globalization: Five Propositions* (PDF online), International Institute for Sustainable Development, 2007, http://www.iisd.org/pdf/2007/trade_environment_globalization.pdf (accessed November 27, 2009).

Make sure all your references—and particularly web sources—are credible.

Websites and web pages

If the resource fits into one of the preceding categories, modify the reference format to reflect the online location. Otherwise use these examples as models for formatting references to general online sources:

27. Environment Canada, *Canadian Climate Normals or Averages 1971–2000* (online), 2009, http://www.climate.weatheroffice.ec.gc.ca/climate_normals/index_e.html (accessed November 27, 2009).
28. Time and Date AS, *The World Clock—Time Zones*, 2009, http://www.timeanddate.com/worldclock (accessed November 27, 2009).
29. *The Cook Islands*, undated, http://www.ck (accessed November 27, 2009).

Electronic books

30. C.A. Bowers, *Transforming Environmental Education: Making the Cultural and Environmental Commons the Focus of Educational Reform* (online), Ecojustice Press, no location specified, 2006, http://hdl.handle.net/10535/15 (accessed November 27, 2009).

31. Translation Bureau, *The Canadian Style: A Guide to Editing and Writing* (online), Dundurn Press, Toronto, 1997, http://site.ebrary.com.library.smu.ca:2048/lib/smucanada/docDetail.action?docID=10221229 (accessed November 27, 2009).

Articles in electronic journals

32. D.R. Caldwell, Unlocking the mysteries of the bounding box, *Coordinates* (online), Series A, no. 2, 2005, http://purl.oclc.org/coordinates/a2.htm (accessed November 27, 2009).
33. A. Morrison, The geospatial content of public transport websites for 60 localities worldwide, *Journal of Maps* (online), vol. 2007, pp. 1–19, 2007, 10.4113/jom.2007.76 (accessed November 27, 2009).

Email, discussion lists, and newsgroups

Recall that it may be prudent and courteous to obtain permission from the sender before using these materials as sources.

Format information as consistently as possible with other references in the list.

34. K. Thomson, <k.thomson@massive.com.au>, Historical Geographies of South Australia (email), IAG-list, <iag-list@ssn.flinders.edu.au>, April 9, 2005.
35. W. Brown, <william.brown@hotmail.com>, Job opening here (email), Personal email to Melanie Smith, <m.smith@msn.com>, June 3, 2006.

Second and subsequent note references to the same source

Second and subsequent references to a source do not need to be as comprehensive as the initial reference. They must simply provide the reader with an unambiguous indication of the source of the material.

1. B. Mitchell, *Resource and Environmental Management* (Toronto: Pearson Education Canada, 2001), p. 56.

Subsequent references:

2. Mitchell, *Resource and Environmental Management*, p. 12.

Another example:

1. D. Beeby, "Report says Fisheries still ignoring environmental rules," *New Brunswick Telegraph-Journal*, Fredericton, New Brunswick, November 24, 2003, p. D7.

Subsequent references:

2. Beeby, "Report says Fisheries still ignoring environmental rules," p. D7.

Bibliography

The bibliography at the end of a work using the note system of referencing includes all the works consulted irrespective of whether they are cited in the text. Bibliographies are arranged in alphabetical order by authors' surnames to make it easier for the reader to find the full details of sources you have cited in the text. A bibliography contains the same information about the works cited as the footnotes or endnotes, but, as in a reference list, the author's surname appears first (rather than his or her initials, which appear first in the notes).

Substantive notes

Sometimes you may wish to let your reader know more about a matter discussed in your essay or report, but you believe the extra information is peripheral to your central message. You can put this information in either an endnote or a footnote. This kind of note is called a *substantive note*.

If you are using the numerical system of referencing, substantive notes should be incorporated into the sequence of your references and should not stand apart from the references. If you are using the author–date system, these notes will be placed either at the bottom of the relevant pages or at the end of the document (before the list of references) in a separate section headed "Notes." Sometimes symbols (e.g., *, §, ¶, ‡) are used for these notes, but superscript numerals (e.g., [1, 2]) are preferable.

Avoiding Biased Language and Stereotyping

You may remember from the chapter on essays that through writing we can shape the world in which we live. By using **discriminatory language** that contains sexual or other biases and stereotypes in our writing and speaking, we may, therefore, unwittingly be contributing to those unacceptable forms of discrimination in society. Biased language "makes unnecessary distinctions about gender, race, age, economic class, sexual orientation, religion, politics, or any other personal information that [is] not necessary to a text's argument or intent" (Writing Center, Wilkes University, undated, unpaginated). Stereotyping involves the use of "words, images, and situations that reinforce erroneous preconceptions or suggest that all or most members of a . . . group have the same stereotypical characteristics" (Translation Bureau, 1997, p. 253).

This section includes ideas, information, and examples that have been synthesized from the following sources: Translation Bureau (1997), Goldbort (2006), Montello and Sutton (2006), Driscoll and Brizee (2012), and the Writing Center, Wilkes University (undated).

Sexual bias and stereotyping

Language can be sexually biased or contain stereotyping in various ways. To avoid biased or stereotypical language, follow these guidelines:

- Avoid using a single-sex word (masculine words such as *he*, *him*, and *his*) when the sex of the person referred to could be male or female. Look for ways to rewrite sentences so that plural terms can replace inappropriate singular terms (e.g., *their*, not *his*). (Remember, though, that it is grammatically incorrect to use a plural pronoun, such as *they*, to refer to a singular noun.) Another solution is to say "he or she" or "his or her" from time to time—but not too often.
- Avoid using words that refer to one sex when both are being discussed (e.g., "man's impact on the environment" instead of "human impact on the environment").
- Avoid using words that encompass women *and* men when reference is being made to one sex only (e.g., using *parents* when only mothers are actually being discussed).
- Avoid using terms that trivialize or denigrate a person's role or contribution (e.g., referring to an "office girl" instead of an "assistant").
- Avoid using terms to characterize men or women that emphasize stereotypical characteristics (e.g., men depicted as proficient drivers, women depicted as incompetent or careless drivers).
- Avoid treating men and women differently in situations that are similar (e.g., a woman's appearance or clothing is described in a report of an event even though those details are not pertinent to the story, while similar details are not given for a man in a parallel situation).

Reflect carefully on the language you are using to make sure it is not discriminatory.

Other biases and stereotyping

In addition to sexual biases and stereotyping, discriminatory treatment occurs when people are judged on the basis of their race, ethnicity, nationality, or sexual orientation rather than on their individual merits or character. Discriminatory treatment has its basis in a dichotomy between the writer's perception of an "in-group" and an "out-group"; that is, the former group is favoured by the writer over the latter. This does not mean that the in- and out-groups both have to be mentioned explicitly for a biased or stereotyped view to be transmitted. Favourable characterization of the in-group or demeaning characterization of the out-group often provide evidence of an author's biases solely by the language used.

As you prepare an essay, report, or presentation, try to avoid terminology and ideas that are biased or contain stereotypes. When you have completed your work, read through it, ensuring that your language does not discriminate against certain people.

Salient Notes on Punctuation and Spelling

How can something as small as a single comma make all the difference in writing? Consider the main title of a book about the proper use of punctuation by Lynne Truss (2004): *Eats, Shoots & Leaves*. The title refers to a joke (about a panda bear who walks into a bar, orders a sandwich and eats it, then produces a gun and opens fire) with a punchline that depends on a poorly punctuated description of what pandas eat. The presence or absence of a comma completely changes the meaning:

A panda bear eats, shoots and leaves.

A panda bear eats shoots and leaves.

A comma changes the meaning of *shoots* and *leaves* from action verbs (as well as eating, what does the panda bear do? It shoots [a gun] and leaves [the room]) to direct objects (what does the panda bear eat? [Bamboo] shoots and leaves). This example nicely illustrates how careful attention to punctuation placement and use can have critical consequences for meaning and should, therefore, be an important consideration when you write. As shown at the end of this section, paying attention to correct spelling is a similar concern.

One of the most important skills you should have by the time you have completed your university degree is the ability to communicate clearly. The following notes are intended to help rectify some of the most common punctuation problems of written English. Take the time to read and absorb them. If you are still unsure about certain technical aspects of writing, consult more extensive guides dedicated to explaining the rules of grammar and punctuation (e.g., Turabian, 2013). Commit yourself to improving your writing by paying attention to punctuation during drafting, revision, and proofreading stages of your work. Also take note of Box 6.3, which lists some common spelling errors that occur because two words, spelled differently, sound alike.

Box 6.3 ◉ Common spelling "mistakes": Right spelling, wrong word

Have you ever written "could of" when you should have written "could have"? (e.g., "It could of been important"—incorrect!). If you say this phrase quickly and without clear enunciation, you might not detect much difference. How often do you see "I'm sorry to hear *your* not feeling well" (which should be "I'm sorry to hear *you're* [*you are*] not feeling well")? Identify the errors in these sentences:

> Do you know what time their open?
> Russia is a larger country then Canada in both population and area.

Some of the most common "spelling" mistakes are not actually misspellings but, rather, the use of a wrong word for the situation. These errors often occur because two words are *homophones*, meaning they *sound alike* but differ in meaning. Writers need to pay careful attention to the meaning and proper use of their words in conjunction with the actual spelling of those words.

Familiarize yourself with the proper use and meaning of the words in the following groups, not just their sounds. Be aware of these commonly confused groups of words. To emphasize, all of these words are spelled correctly, but often the wrong word is used for the given context.

- accept/except
- affect/effect
- its/it's
- of/have
- principal/principle
- than/then
- their/there/they're
- to/too/two
- your/you're

The list above presents some common "sounds like" spelling substitutions, but remember not to neglect accuracy in spelling generally. Proofread your final draft for spelling errors before you submit it. If you need a list of correct spellings, consult the ultimate word list—a dictionary (in particular, a Canadian dictionary)! Word-processing programs also have spell-checking tools as standard features. Use them to improve your writing, but be sure the program is configured to use Canadian spelling.

Comma [,]

- Separates principal clauses in a compound sentence. (A principal clause has a subject and verb. A compound sentence consists of two or more principal clauses, which are joined by *and*, *but*, *or*, or *for*).

The battered yellow coaches were filled to overflowing, and on the curves I could see people on the roofs of the following carriages.

The aim was to examine sustainability, but the experiment failed.

- Follows a subordinate clause or a long phrase at the beginning of a sentence.

  After the recommendations were implemented, further evaluations were conducted.

- As part of a pair, brackets or separates parenthetical information in a sentence.

  The most common, and most easily rectified, problems on essay writing emerge from incorrect acknowledgement of sources.

- Separates the items in a list.

  The equipment included one inflatable boat, one motor vehicle, and a helicopter.

The comma before *and* is optional in the previous example, but in certain situations including a comma in that position will eliminate ambiguity in the meaning of a sentence. Compare the following two sentences, which differ only by the omission or presence of one comma:

Dr Martin went to dinner with two of his former students, Ms Jones and Mr Clark.

Dr Martin went to dinner with two of his former students, Ms Jones, and Mr Clark.

How many people were in the dinner party, three or five? Inclusion of the comma creates a different meaning. The first sentence leads us to assume that Dr Martin had dinner with two people, Ms Jones and Mr Clark, who are his former students. The second sentence indicates that Dr Martin was joined for dinner by four people: Ms Jones, Mr Clark, and two others who are his former students.

Period [.]

- Ends a complete sentence.

  Geography is about more than just names on maps.

- Follows an abbreviation.

  p. (for *page*), ed. (for *editor*)

Ellipsis points [. . .]

- Show that words have been left out of a quotation.

  As the report claims, "there are many factors determining the state of the physical environment . . . but the most important is human intervention" (Murray, 2009, p. 46).

- In *informal* writing may signify that there is more of what you are saying but that you are not putting it on the page.

  She went on and on about the rate of inflation, share-market movements, currency exchange rates . . .

  This construction is not acceptable in formal academic writing.

Semicolon [;]

- Connects two sentences or main clauses that are closely connected but are not joined by a conjunction.

  The initial survey revealed significant interest; results showed that further action is appropriate.

- Separates complex or long items in a list, especially if any of the items in the list contain commas.

  The following factors are crucial: the environmental impact statement; the government and union policies; the approval of business and council; and public opinion.

  Commas are also used in lists of simpler items.

Colon [:]

- Precedes a list that is not a grammatical part of the introduction to the list.

  The following factors are important: precipitation, temperature, and population.

  There are five Great Lakes: Superior, Michigan, Huron, Erie, and Ontario.

  Do *not* use a colon if the list is a grammatical part of the sentence.

  The most important factors are precipitation, temperature, and population.

The five Great Lakes are Superior, Michigan, Huron, Erie, and Ontario.

- Precedes a long quotation introduced by words like *writes*, *says*, *argues*.

  Openshaw (1999, p. 81) writes: "the fundamental technical change that is underpinning the development of the new post-industrial society is the transformation of knowledge which can be exchanged, owned, manipulated and traded."

Quotation marks (double or single) [" or ']

- Indicate that a word, phrase, sentence, or group of sentences is a quotation. (If a long quotation is set off from the rest of the text as a "block quotation," quotation marks are not used.)

  "For our purposes, militarism can be broadly defined as 'a set of attitudes and social practices which regards war and the preparation for war as a normal and desirable social activity.' " (Mann, 1988, p. 166)

- Are used around the titles of journal articles, chapters in a book, and so on.

  Harvey's (1990) paper "Between space and time" is an example of an important contribution to the field.

Double quotation marks are used more frequently than single quotation marks, but if there is a quote within a quote, then the inner quote would be enclosed within single quotation marks.

Apostrophe [']

- Shows that a letter (or sequence of letters) is missing, as in contractions (e.g., *I'm*, *we'll*, *can't*). Note that contractions belong to informal writing and are not usually acceptable in formal academic writing.
- Indicates possession, as follows:
  - If the noun is singular, add an apostrophe plus s to the end of the word.

    The researcher's results [that is, the results of one researcher]; Burdess's new book [the new book by Burdess]

  - If the noun is plural and ends in s, add an apostrophe to the end of the word.

    The researchers' results [that is, the results of more than one researcher]; the Harrises' experiment [an experiment by two or more people called Harris]

- If the noun is plural and does not end in s, add an apostrophe plus s.

  children's games [games of children]; women's answers [answers of women]

Apostrophes are one of the most commonly misused or neglected elements of punctuation.

- Avoid using an apostrophe plus s to show possession with inanimate objects.

  The findings of the study [*not* the study's findings].

- Note that *it's* means "it is." The possessive of *it* is *its* without an apostrophe.

  The cat is chasing its tail [*not* The cat is chasing it's tail].

Upper case letters

- In titles and headings, do *not* use upper case for articles (*the*, *a*, *an*), coordinate conjunctions (*and*, *but*, *for*, *or*, *nor*), or prepositions unless at the beginning.
- Use capital letters only for proper nouns, that is, the names of particular persons, places, or things: Peter, France, English, Cree, Ottawa River, Main Street, Mount Robson.
- Within sentences in the body of a written work, do not use upper case for a word just because it is a keyword for the topic, such as "drumlins" in an essay about sub-glacial landforms.

Numerals

- Spell out numbers under ten (e.g., *four* but *16*). But always use figures with units of measurement and symbols, such as %: e.g., "9 mm" but "nine field sites"; "5%" but "five books"
- Do not use figures at the beginning of a sentence: e.g., "900 workers were laid off." Sentence should always begin with a word. You should write: "Nine hundred workers were laid off."
- Put a space between the numeral and the unit of measurement, and do not use periods after units of measurement: "40 kg of the sample", *not* "40kg of the sample" or "40 kg. of the sample."

Further Reading

Anglia Ruskin University. (undated). *Harvard system of referencing guide.* http://libweb.anglia.ac.uk/referencing/harvard.htm (accessed 18 November 2013).

- This is a comprehensive list of reference style examples in the author–date (Harvard) system.

Eichler, M. (1991). *Nonsexist research methods: A practical guide*. London: Routledge.

- This book provides a comprehensive explanation of the ways in which research practices can be sexist. It also includes a discussion of sexism in language (Chapter 7).

Library of Congress. (undated). *Citing primary sources*. www.loc.gov/teachers/usingprimarysources/citing.html (accessed 17 November 2013).

- This page focuses specifically on electronic sources. Examples are given in two common reference styles employed more commonly in the humanities and some social science publications: MLA style and Chicago style.

Purdue OWL. (2013). *Spelling: Common words that sound alike*. http://owl.english.purdue.edu/owl/resource/660/01 (accessed 20 November 2013).

- Spelling is just one topic about writing available at Purdue University's Online Writing Lab. (This resource is just one of many that provide help with writing, both online and in print form. See the Further Reading lists in other chapters for more suggestions.)

Translation Bureau. (1997). *The Canadian style: A guide to editing and writing* (2nd ed.). Toronto: Dundurn Press.

- This is the Canadian government's style manual and is suitable for all Canadian writers seeking a standard guide to follow. It includes a chapter on eliminating stereotyping in writing. Note that its reference format gives the author's full first name and not just the initial.

Truss, L. (2004). *Eats, shoots & leaves: The zero tolerance approach to punctuation*. New York: Gotham Books.

- This book is a lively and informative treatise on the importance of using punctuation correctly.

Turabian, K. (2013). *A manual for writers of research papers, theses, and dissertations* (8th ed.). Revised by Booth, W.C., Colomb, G.G., Williams, J.M., and the University of Chicago Press editorial staff. Chicago: University of Chicago Press.

- An extremely detailed guide to writing in the "Chicago style," this book would be useful for questions on formatting, citations, and references.

7 Making a Poster

A picture can be worth a thousand words—
but only if a reader can decipher it.

—Stephen Kosslyn

Key Topics

- Why make a poster?
- How are posters produced?
- What are markers looking for in a poster?
- Guidelines for preparing a good poster

Posters are a useful way of presenting the results of research and other scholarly enterprises. They are an effective and swift means of presenting an idea or set of ideas, and they are finding increasing use at professional conferences and other gatherings (Pechenik, 2007, p. 244). Producing a good poster can be challenging, however, because you need to provide an effective combination of graphic and written communication. This chapter outlines some of the keys to effective poster production. The discussion covers layout, visibility, and the use of colour and type style. While there is also a brief reference to the use of illustrative material, extended discussions on the production of figures, tables, and maps can be found in chapters 8 and 9.

Why Make a Poster?

A poster presents an argument or explanation, summarizes an issue, or outlines the results of some piece of research in succinct visual form. Posters are especially good for promoting informal discussion (Morgan and Whitener, 2006, p. 105) and for showing results that require more time for interpretation than is possible in, say, an oral presentation. However, they are not as useful for reviewing past research or presenting the results of "textual" research (Lethbridge, 1991, p. 14). Students with little experience in presenting their work may find a poster presentation to be less threatening than an oral presentation (Anholt, 2006, p. 108). In contrast to an oral presentation that an audience may hear only once, a poster can also be left on display, potentially increasing the size of your audience.

Your professor may ask you to prepare a poster for a variety of reasons (see Hay and Thomas, 1999, and Knight and Parsons, 2003, for discussions of this topic). Posters

- add variety and new challenges to a course;
- encourage the expression of complex information and ideas through careful combinations of text and graphical elements;
- develop and test skills in graphic communication;
- stimulate critical thought; and
- offer the prospect of encouraging student–staff interaction.

However, making a formal poster is a relatively uncommon exercise in lower-level undergraduate courses. It becomes more common at advanced levels, such as when a student participates in a conference or forum to present results of a larger research project. Posters are also used extensively by graduate students and professionals at conferences and exhibitions as an alternative to an oral presentation. Consider also the arguments made in Chapter 11 about the need for geographers and environmental scientists to communicate with the public, often outside the traditional academic setting. A poster (or an oral presentation or media release) might be the best method of achieving the desired public engagement.

Posters are a challenging means of communicating information. They require you to express complex ideas with brevity and grace, and to balance text with graphics of high quality. Alley (2003) emphasizes that posters are not just report pages or presentation slides attached to a board; instead, "posters, when effectively designed, are a medium distinct in typography, layout, and style" (p. 211).

Examples of posters can be found on the web. The website of Hess et al. (2013) presents several examples and discusses the positive and negative aspects of each one. Alley (undated) and Purrington (undated) also give advice and poster examples. When viewing the examples, be sure to evaluate such aspects as organization and clarity, and ask yourself if the most important material has been included.

How Are Posters Produced?

Physically, a poster may be up to about 1.5 metres wide and one metre high, although the maximum dimensions will often be determined by the size of display boards or spaces. There are two main techniques for producing posters, one of which (cut-and-paste collage) is largely outdated as the other (large-format printing) becomes more widespread. If you do not have access to a large-format plotter or a printer capable of printing at poster size in one pass, you can construct a cut-and-paste poster by preparing smaller individual elements or sections—such as titles and headings, graphic materials, and blocks of text—and physically attaching them to a single background. Often the background is a piece of stiff cardboard, which is unsuitable for folding or rolling and therefore can be difficult to transport. To improve the poster's appearance, you should prepare individual elements on a computer and print them out. Printing by hand will make your poster resemble an elementary-school craft project. At the university level, you should be striving for a higher level of presentation.

With the large-format printing technique, a poster is created in a software program (such as Adobe Illustrator or Microsoft PowerPoint), with the individual elements brought together digitally before the full poster is printed in one pass. Cutting and pasting is still done, but electronically rather than manually. Large-format printers use rolls of paper that are about one metre wide, thus defining the maximum size of one dimension of a poster. The potential length or width of the poster is unlimited but the width will usually not be greater than twice the height. At professional conferences, posters produced on large-format printers are now almost exclusively the norm as access to such printers has increased and printing costs have become reasonable. Presenters and viewers agree that large-format posters have a more professional look than cut-and-paste collage. Moreover, these posters can be rolled up without being damaged, which makes them easier to transport.

However, to use large-format printing, you must have access to a suitable software program as well as proficiency in creating and manipulating digital materials. For lower-level courses in particular, you may be permitted, or required, to create a cut-and-paste poster. For upper-level courses, or if you are attending a conference, you will likely be expected to create a large-format poster. If you do not have access to a large-format printer or are directed to create a cut-and-paste poster by your professor, the principles of good poster design still apply. In the rest of this chapter, little distinction is made between cut-and-paste and large-format printed posters.

What Are Your Markers Looking for in a Poster?

Poster production is not an explicit test of your artistic abilities, but if you have such aptitudes you should take advantage of them. Regardless of your artistic prowess, however, a few principles are critical to creating a successful poster (Box 7.1). Not surprisingly, these principles are similar to the keys that unlock successful written and oral communication.

Posters are sometimes produced for audiences other than your professor and your classmates. For example, you might be asked to make a public-information poster on strategies for making the best use of green waste. If you have such an assignment, always consider your target audience and ensure that your language, ideas, and choice of graphics suit that audience. Perhaps more commonly, though, you will be working on "academic" posters. In addition to the principles of poster production described above, these may require that you

Academic posters are quite different from promotional posters.

- express a problem and resolve it;
- argue or explain an issue; and/or
- collect and evaluate evidence about a topic.

While a completed poster should require no additional information for the viewer to be able to understand the content, there may be occasions when your professor will ask you to stand by your poster and

A good poster should stand alone comprehensibly, requiring no further explanation.

Box 7.1 ◉ Principles of poster production

The following principles should guide your poster production. Make sure that your poster is

- *eye-catching*. The poster should make a good first impression. It should grab the viewer's attention through layout, colour, title, and other devices.
- *brief*. The poster should make its points quickly.
- *coherent*. The poster should make a logical, unified statement requiring no further explanation. It should be comprehensible to the intended audience and must be capable of standing alone.
- *clear*. The poster should follow a clear path through the subject matter. Keep the poster simple. Keep it focused. Over-complicated posters discourage and confuse the viewer.
- *based on evidence*. The poster should present an argument that is supported by accurate, referenced evidence.

answer questions about it—as is standard practice at academic and professional conferences.

Your poster should reflect *critical thinking* rather than simply your capacity to *describe* some phenomenon. If you are given a free choice of poster topic by your professor, you would be wise to check that the topic you have selected is appropriate before you proceed too far. Make sure that you are satisfying the intellectual demands of the exercise as well as the graphic requirements. The poster assessment form in Box 7.2 describes some general criteria for assessing posters. As you work on your poster, you should assess it continually against those criteria. This process should help you produce a high-quality piece of work.

Box 7.2 ◉ Poster assessment form

Student Name: **Grade:** **Assessed by:**

The following is an itemized rating scale of various aspects of written assignment performance. Sections left blank are not relevant to the attached assignment. Some aspects are more important than others, so there is no mathematical formula connecting the scatter of ticks with the final grade for the assignment. Ticks in either of the two boxes left of centre mean that the statement is true to a greater (outer left) or lesser (inner left) extent. The same principle applies to the right-hand boxes. If you have any questions about the individual scales, comments, final grade, or other aspects of this assignment, please see the assessor indicated above.

Quality of argument					
Question or relationship being investigated is stated clearly					Ambiguous or unclear statement of purpose
Poster fully addresses the question					Poster fails to address the question posed or issue raised
Poster "stands alone" and requires no additional explanation					Poster is difficult or impossible to comprehend without additional information
Logical and adequate explanation of the issue under investigation					Illogical or inadequate explanation of the issue under investigation
All components in presentation are given appropriate level of attention					Treatment of components is insufficient or unbalanced

Continued

Quality of evidence

Argument is well supported by evidence and examples					Supporting evidence or examples are inadequate
Presentation of evidence and examples is accurate					Evidence is incomplete or questionable

Use of supplementary material

Use of figures, tables, and other illustrative material is effective					Illustrative material is not used when needed or not mentioned in text
Illustrations are presented correctly					Illustrations are presented incorrectly

Poster appearance

Poster is produced carefully					Presentation is sloppy
All text is legible from 1.5 m					Text is not legible from 1.5 m

Sources and referencing

Number of sources is adequate					Number of sources is inadequate
Acknowledgement of sources is adequate					Acknowledgement of sources is inadequate
Citations are correct and consistent					Citations are incorrect or inconsistent
Reference list is presented correctly					Reference list contains errors or inconsistencies

Assessor's comments

___

___

___

___

Guidelines for Designing a Poster

Preliminary layout

Because thoughtful layout and design are such important parts of effective communication, you should reserve a big chunk of time for them when you are making a poster. Work that may have taken many hours to prepare can be ruined by an ill-conceived layout. Even while you are conducting the research for the exercise, you should consider how the information you generate will contribute to your poster's effective design. Produce sketches or mock-ups of poster layouts before deciding on the final set-up. Discuss these sketches with others who will bring their fresh, critical outlook to your work. Following these steps will likely save you time in the long run and help you to make sense of the issue under discussion in your poster. For example, in creating a preliminary layout of the poster, you may find weaknesses and gaps in any argument you are developing or in any relationships you are exploring. These discoveries should prompt you to undertake additional investigation that will contribute to the production of a better project. Of course, this means that you cannot leave poster production to the night before the assignment is due, no matter how straightforward the project may appear initially!

Organization

Careful organization, or composition, is crucial to effective poster communication. Your poster should have a logical and clearly apparent structure (see Figure 7.1[a]). Besides graphics, most posters have seven important components:

1. *Title section*. This should be meaningful, visible, brief, and memorable. The title section should normally also include the name(s), affiliation(s), and contact information of the poster's author(s).
2. *Abstract*. This should summarize the poster's content. Detailed information about writing abstracts is given in Chapter 4.
3. *Introduction*. Here, you provide a short statement of the topic investigated as well as the project's significance.
4. *Body*. If your poster is presenting the results of some research, this section might be split into a number of parts. The first might be *materials and methods*, where you explain, in sufficient detail for any readers to determine the scope of your study, the research techniques used, and the validity of the data obtained. The second part of the body is likely

to be a statement of *results*. This is a vital part of the poster. It enables readers to examine the data on which your conclusions are based and to evaluate their validity. Most readers need to see the information on which conclusions are based before they are able to accept them.

5. *Statement of conclusions* and/or *directions for future research*. The conclusions let readers know how you have interpreted the information presented in the poster. You might also explain how future work could follow up on some particular aspect of your findings.
6. *List of references*. Just as with an essay or report, you must specify the source of any information that is not your own.
7. *Acknowledgements*. This is an opportunity to thank people who assisted in your research or with the production of the poster. Limit the acknowledgements to those who made a direct contribution to the work on display.

Evidently, then, posters are little different structurally from essays and other forms of expression. However, unlike written forms of communication, posters do not have to be set up in a linear form, with readers moving from top left to bottom right. *Spider* diagrams (Figure 7.1[b]) show how various factors contribute to an understanding of the central issue (e.g., what are the factors that determine social power in a community?). *Cyclical* diagrams are useful when the phenomenon being investigated has no clear beginning or end: for example, the ocean food chain, the hydrological cycle, or the cycle of poverty. However, while posters do offer this flexibility of presentation, you still have to provide your readers with a clear sense of direction. To this end, you must usually guide readers through the sections of the poster with informative headings, section numbers, or arrows. Readers do not know their way around your poster as you do. Lead them along. Be aware, however, that after reading the title and scanning the poster, many people move straight to the conclusions and make a decision whether to spend more time reading the poster or to move on to the next one (Beins and Beins, 2008). If the conclusions seem important, the viewer may then decide to read the rest of the poster. Therefore, give some hard thought to your title and to the way you present your conclusions.

> ***Make sure the structure of your poster is clear to your audience.***

Cut-and-paste posters may be enhanced by making them interactive (Vujakovic, 1995, p. 254). Readers might be asked questions to answers that can be revealed by lifting flaps or overlays, or by turning circles of cardboard. Devices such as these will almost certainly encourage your audience

(a) Structure based on progressive sections

Title	*Author, Affiliation, and Contact Information*	
Abstract	*Results—Year A*	*Conclusions*
Introduction		*Further Research*
Materials and Methods	*Results—Year B*	*List of References*
		Acknowledgements

(b) Structure based on a spider diagram

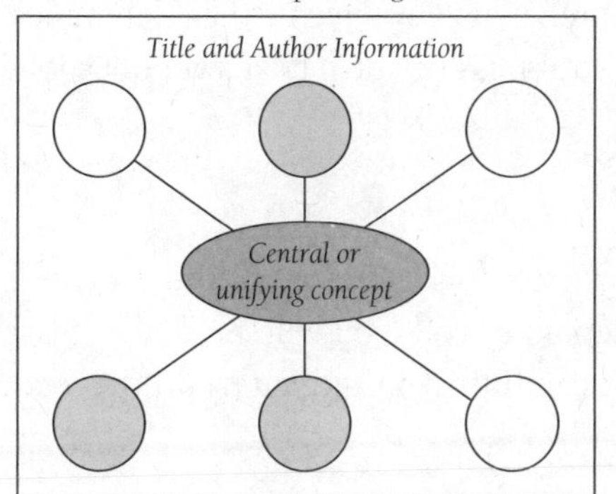

(c) Structure based on a cyclical diagram

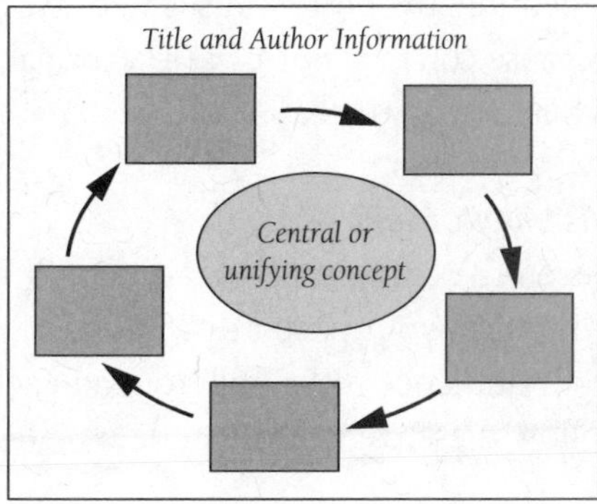

Figure 7.1 ◉ Examples of different poster structures

Source: Hay and Thomas (1999, p. 211)]

to interact with the poster. They should be kept uncomplicated and sturdy, however, as you do not want an interactive element to fall apart after one person has used it.

Text

Number of words and level of detail

Unlike many promotional or advertising posters, academic posters almost always contain some text. Together with graphics, text contributes to the introduction, explanation, and discussion of your work. You should, however, limit the amount of text. Although a poster is physically large, Lethbridge (1991, p. 18) makes the point that a poster will be effective if it uses only 250 to 400 words. Confine the text you do use to short sections that complement the illustrative components. Make sure no single block of text takes longer than 30 seconds to read (Knight

> ***Use small amounts of text on your poster and be sure the text is visible from at least 1.5 metres.***

and Parsons, 2003, p. 155); poster viewers do not enjoy reading long passages of text. Use point-form statements rather than full paragraphs. Ineffective posters are often those that contain too much text or text that is presented in a small number of dense blocks.

In a poster, it is not necessary to present all of a project's intricate details. In much the same way as an oral presentation highlights points that can later be explored by interested members of the audience, so too a poster can be used to present the most important elements of your work. Onlookers whose interest has been aroused can then speak to you about details in a face-to-face conversation (Nicol and Pexman, 2003). That said, an earlier point bears repetition: the poster must still be comprehensible *without* further elaboration because you may not always be available to stand beside it to answer questions while it is on display.

Visibility of text

Given the amount of work you will have to do to get the number of words you use in your poster down to an acceptable level, you will, no doubt, want your audience to be able to read each carefully selected letter! To this end, follow these suggestions:

- All materials on your poster should be clearly legible from a distance of about 1.5 metres.
- Titles and headings should be readable several metres away (so as to draw people in to take a closer look).
- Use upper and lower case letters throughout the poster text. Do not use only capital letters. UPPER CASE IS A LOT MORE DIFFICULT TO READ QUICKLY.
- Confine textual material to a number of brief sections. Do not write an essay and paste it into the poster! Few people, if any, will read it.

Use larger fonts for headings that are higher in the hierarchy and reduce the font size for lower-level headings (Box 7.3). All headings at the same hierarchical level should be in the same typeface. The common type size for essays and reports is much too small for the body of poster text. If you need to use 10 or 12 point type in order to fit all your text on the poster, the poster probably contains too much text. The solution to this problem is to abbreviate and rewrite the text.

Type is one of the most important aspects of visual design, especially for headings, and you should use a typeface that is suited to the subject material. Whereas the body of the text needs to be clear and simple, you may be able

Box 7.3 ◉ Suggested type sizes for posters

• First-level headings	96–180 point	(27–48.5 mm)
• Second-level headings	48–84 point	(12.9–25.4 mm)
• Third-level headings	24–36 point	(5.9–8.7 mm)
• Text and captions	14–18 point	(3.2–4.6 mm)

to add extra graphic character to the poster through your choice of a suitable type style. Type characteristics (see Box 7.4) play an important role in attracting attention, and they contribute to the poster's overall theme and clarity.

For example, a very modern typeface might be appropriate for a poster investigating the spatial distribution of manufacturers linked to the space industry, and a playbill typeface—first used in the Victorian era on posters advertising stage productions—could be appropriate for a poster considering nineteenth-century Prairie health care.

Box 7.4 ◉ Characteristics of type

- *Face:* the particular character of the letter forms, from which there are hundreds to choose (e.g., Helvetica, Times Roman, Arial)
- *Weight:* the thickness of the letter (regular, **bold**)
- *Style*: roman (upright) or *italic* (slanted)
- *Size:* height, as measured in points, such as 14 point
- *Colour:* even with only a black-and-white printer, may be **black**, white, or grey scale

Colour

Often, one of the most striking and emotive elements of a poster is colour, which can add to or detract from the overall impact of the project. Colour can, moreover, attract attention and improve the presentation of results (Nicol and Pexman, 2003), and it can highlight important elements of a poster or suppress less important ones.

> ***Use type characteristics and colours that complement your material.***

Although colour is important, you should use it judiciously. To avoid confusion and chaos in the poster, use as few colours as possible. Pechenik (2007, p. 249) suggests that a single background colour should be used to unify the presentation. Take care too in your choice of combinations of text and background colours. There should always be enough contrast between text and background to allow the text to be read easily from a distance. For example, orange text on a yellow background can be difficult to read, as can red text on a green background.

A popular technique is to use a related image as the poster background rather than a solid colour, but if text is placed directly on the image, parts of it can be difficult to read because of a lack of contrast. It can also be frustrating to read if the contrast between text and background changes frequently.

Colour can add symbolic connotations and feelings to a poster's message and may also be used to add information to particular graphics. For example, the use of red and black conveys the message that financial results represent profit and loss statistics, not simply dollar figures.

Tables, figures, and photographs

Tables

Tables on posters should be simple and clear. Readers are unlikely to spend much time trying to decipher complex data sets. In many cases it is better to summarize and depict tabular information in the form of histograms, pie charts, and other graphic devices (see Figure 7.2). These will usually convey your messages faster and more memorably than text alone, and for some audiences tables may be the most appropriate communication device. Although tables are a useful means of presenting precise numerical information, ask yourself whether the data could be transformed into a "picture" as opposed to a table. For example, a table of population figures for Toronto since 1900 might show the changes over time better when presented as a **line graph**. Try to achieve a balance between the use of tables and figures on your poster.

> ***On your poster, consider summarizing data in a figure.***

Figures and photographs

Images are a particularly important part of a poster. For example, you can use photographs to show a forested landscape before and after clear-cutting or to illustrate transformations in a city after a large industry shuts down. You can use accompanying diagrams to explain processes and procedures more effectively than in blocks of text. The web is a vast store of images. A search may reveal helpful illustrations and photographs that may be used *subject to copyright restrictions*. Photographs and graphic images are considered to be someone's work, so even if there is no copyright notice the source must be acknowledged. In addition to non-specific search functions (e.g., Google Images), there are growing numbers of specialist online image galleries that can prove helpful (e.g., the University of Chicago Library's American Environmental Photographs Collection, or National Geographic's website). Given its importance, any graphic material in your poster must be bold *and relevant*. Avoid filling up poster space with pictures that may be attractive but are unnecessary for your purpose as they are likely to detract from the central matters you wish to communicate. Pictures that are merely space fillers suggest a lack of effort.

The web provides a vast array of potentially useful images for posters.

In the process of considering the production of figures for a poster, you should ask several questions:

- What type of "picture" will best illustrate the point? For example, would a pie chart be more effective than a bar chart?
- Which symbols and colours will make a greater impression and will communicate the idea more clearly? For example, would it be effective to illustrate Canada's federal budget deficits in a **bar graph** depicting piles of coins shaded red or black depending on each year's deficit or surplus? The graphical devices you employ may be more readily understood, and may therefore be more effective, if they incorporate symbols and colours that have evolved through tradition, convention, and public recognition to be representative of their content.
- Are all the illustrative conventions likely to be comprehensible to the audience? Maps, for example, should always have a **key** or **legend** explaining all the symbols used.

Figure 7.2 shows an example of a set of numerical data that might be incorporated into a poster display in different forms. In general, you should find a simplified but graphically interesting way of displaying data when producing a poster. Whereas a table (Figure 7.2a) might provide the most accurate record of research data collected, it does not enable the viewer to absorb the major trends as quickly as a graphical depiction would. A graph like Figure 7.2b would allow the reader not only to obtain a picture of the relative trends quickly, but also to make a reasonable interpretation of the data. However, a graph such as this can be modified for a poster display. A simplified and more interesting portrayal, like Figure 7.2c, enables the viewer to grasp the essential pattern quickly. In this example, interest and clarity are achieved by differentiating line style (solid vs. broken), font, and

Company Revenue ($ 000)

Year	Canadian Sources	International Sources	Total
2001	5067	1010	6077
2002	5328	3126	8454
2003	6141	2842	8893
2004	7002	6589	13 591
2005	8269	2354	10 623
2006	2103	2413	4516
2007	2025	4201	6226
2008	1567	3368	4835
2009	4824	423	5247
2010	6041	0	6041

(a) *Table of numerical data*

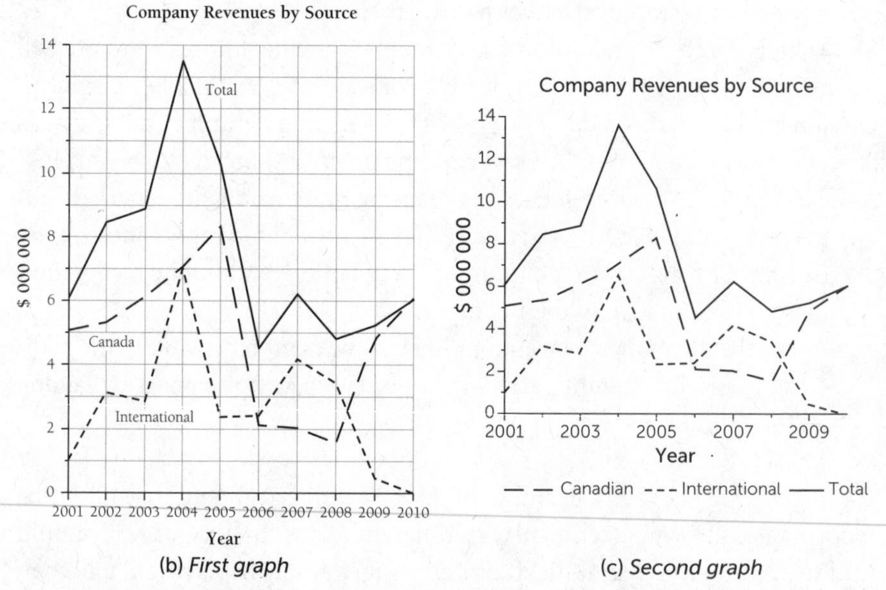

(b) *First graph* (c) *Second graph*

Figure 7.2 ● Converting a numerical table into a poster-ready figure

type size. The addition of colour might also increase a poster's visual appeal. Do not forget to include an appropriate title (and an explanatory **caption** if necessary) for figures.

If you use photographs in your poster, be sure they are of high quality (for example, in focus and with sharp contrast) and large enough to be clearly visible from a distance of one or two metres. If the size of the photographed object would not be immediately clear to your audience, provide some idea of **scale**. For example, include a camera lens cover, a ruler, or a person in the frame when taking the photograph.

> ***If you want to keep your poster, present it at a conference, or display it permanently, consider laminating it.***

You will work long and hard to make a good poster. Protect your investment. If you are feeling especially wealthy and motivated, or want to keep the poster and display it more than once, consider getting it laminated. This will protect it from the weather and from spills or other accidents.

List of sources

As with any other form of academic communication, you must accurately acknowledge any information and ideas, figures and facts, text and tables, or images you have drawn from other sources. These should be cited and listed in exactly the same way as they would be in an essay or report (see Chapter 6 for a discussion of citation and referencing conventions). To save space, the **reference list** may be displayed in a slightly smaller type size than the main text. It is important to acknowledge your sources and to show the bibliographic details on the poster for readers who want to investigate further, but most viewers will not read the reference list as a single block.

Throughout the production of your poster, keep in mind the qualities of a good poster that were introduced in Box 7.1: the poster must be eye-catching, brief, coherent, clear, and based on evidence. Providing, of course, that the poster draws on good research, careful application of these principles should contribute to the production of first-class work.

Further Reading

Alley, M. (undated). *Design of scientific posters.* http://writing.engr.psu.edu/posters.html (accessed 18 November 2013).

- This page is a brief treatment of poster design principles, but it contains useful examples of completed posters and several links to other websites that discuss posters.

Beins, B.C., and Beins, A.M. (2008). *Effective writing in psychology*. Malden, MA: Blackwell.

- Chapter 16 includes advice on presenting during a conference poster session, including paying attention to your personal appearance as well as your poster's appearance. The chapter also provides technical advice on creating a poster with PowerPoint.

Brown, B.S. (1996). Communicate your science! . . . Producing punchy posters. *Trends in Cell Biology*, 6, 37–39.

- This article provides a lively summary of the keys to creating a poster that effectively present your work.

Hay, I., and Miller, R. (1992). Application of a poster exercise in an advanced undergraduate geography course. *Journal of Geography in Higher Education, 16*(2), 199–215.

- Intended essentially for professors, this paper discusses a strategy for incorporating a poster exercise into an upper-level class. Special emphasis is given to the rationale for the project and to the practicalities of poster production, much of which is elaborated upon in this book.

Lethbridge, R. (1991). *Techniques for successful seminars and poster presentations*. Melbourne: Longman Cheshire.

- This short book briefly introduces poster-design principles and goes on to provide a comprehensive discussion of technical aspects of preparing illustrations. Indeed, the title may be a little misleading given the book's emphasis on graphic production techniques.

Pechenik, J.A. (2007). *A short guide to writing about biology* (6th ed.). New York: Pearson Longman.

- Chapter 12, "Writing a poster presentation," offers a useful and detailed discussion of the process of converting a written paper into a good poster presentation.

Purrington, C.B. (undated). *Advice on designing scientific posters*. www.swarthmore.edu/NatSci/cpurrin1/posteradvice.htm (accessed 22 November 2013).

- This is a comprehensive web page that covers poster design and presentation and gives many examples of individual elements of posters as well as full posters.

8 Communicating with Graphs and Tables

As Bertrand Russell once said, most of us can recognize a sparrow, but we'd be hard put to describe its characteristics clearly enough for someone else to recognize one. Far more sensible to show them a picture.

—D. Rowntree

Key Topics

- Why communicate with graphs and tables?
- General guidelines for communicating clearly with graphs
- Different types of graphs
- Preparing good tables

Why Communicate with Graphs and Tables?

In geography and the environmental sciences, words or numbers alone are often not sufficient to convey information effectively. Graphs and tables allow you to display a large amount of information succinctly and help your audience absorb that information readily. Effective graphs and tables can help a reader achieve a rapid understanding of an argument or issue.

Graphs employ human powers of visual perception and pattern recognition (Kosslyn, 2006) that are much better developed than our capacity to uncover meaningful relations in numerical lists (Krohn, 1991, p. 188). Often we see things in graph form that are not apparent in tables or text. Krohn argues that graphs reveal relationships that allow both the numbers upon which the graphs are based and the concepts by which we understand those numbers to be reinterpreted. Thus, graphs are critical interactive sites for comprehending the world around us.

Graphs can reveal relationships that are difficult to see in tables or text.

In addition to their intellectual functions, graphs and tables can enhance various forms of technical writing in different ways (Eisenberg, 1992, p. 81):

- *In essays and reports*. Graphs and tables summarize quantitative information, freeing up text for comments on important features.
- *In instructions*. Graphs and tables may help people to understand the characteristics of a phenomenon or the principles behind the operation of some process.
- *In oral presentations*. Graphs and tables (and other diagrams or illustrations) relieve monotony, help guide the speaker, and aid the audience's understanding of data.

This chapter first discusses the character and construction of graphs, and then of tables.

General Guidelines for Communicating Clearly with Graphs

Good graphs are concise, comprehensible, independent, and referenced.

1. *Concise*. Graphs should present only that information which is relevant to your work and necessary for you to make your point. Critically review the data you are going to illustrate to find out what those data "say" and then let them say it graphically with minimum embellishment (Wainer, 1984, p. 147; Nicol and Pexman, 2003, p. 6). If you reproduce a graph you have found in your research, you may need to redraw it to remove irrelevant details (but remember to cite the original source).
2. *Comprehensible*. Your audience must understand what the graph is about. According to Kosslyn (2006, p. 4), "a graph is successful if the pattern, trend, or comparison it presents can be immediately apprehended." Provide a clear and complete title and caption (answering "what," "where," and "when" questions), a key or legend, and effective labelling. Effective labels are

   - easy to find,
   - legible (make sure the font is appropriate), and
   - easily associated with, or close to, the corresponding axis or object.

   Blank areas within the data region of a graph may be used for labels or annotations if the visibility of the contents is not impaired. In short, the displayed information should be clear and easy to read. If your graph displays two or more data sets, they must be easily distinguished from one another. Graphs should include no more than *four* simultaneous symbols, values, or lines (Cartography Specialty Group of the Association of American Geographers, 1995, p. 5). Furthermore, lines or symbols should all be different enough to be easily distinguished from one another.

You can also make a graph more comprehensible by making effective use of the data region (the part of the graph where the data are displayed). Choose a range of axis scale marks that will allow the full range of data to be included while ensuring that the scale allows the data to fill up as much of the data region as possible. For example, if you are graphing percentage data and the highest individual value is 40 per cent, showing the full scale from 0 per cent to 100 per cent would result in too much empty space. If you take photographs, this principle will be familiar to you. Just as good photos will usually "fill the frame," so too should a good graph fill the data region. Finally, the tick marks on each axis should be frequent enough for a reader to work out accurately the value of individual data points (Pechenik, 2007, p. 170).

3. *Independent.* Graphs should be able to stand alone. Someone who has not read the document associated with the graph should be able to look at the graph and understand its purpose and meaning.
4. *Referenced.* You must acknowledge your sources. Use an accepted reference system to note the sources of graphs you use or of the data in graphs you create. Each graph should be accompanied by summary bibliographic details (author, date, and page in the case of the author-date system) or a note identifier allowing the reader to find out where the graph or the data upon which it is based came from. A reference list at the end of your work should then provide the full bibliographic details of all sources. If the source of your data or graph obtained that information from another source, both sources need to be acknowledged in a citation. For example, a graph portraying the relationship between slope gradient and drainage basin area appearing as Figure 5.1 in Hovius et al. (2004) was originally published by Montgomery (2001). A correct citation for this material is (Montgomery, 2001, in Hovius et al., 2004, Figure 5.1). The citation would not simply be Hovius et al. (2004). To simplify, obtain and cite the Montgomery (2001) source, bypassing the need to cite Hovius et al. (2004).

Computerized versus Manual Production of Graphs

Just as a word-processing program is now the standard way of writing, although one could still choose to write by hand, people now commonly use software programs to produce graphs instead of doing so manually. One such program that can be used to produce adequate graphs for most undergraduate assignments is Microsoft Excel. Using a program will usually produce a graph with at least a minimally acceptable appearance if certain guidelines are followed (see Box 8.1); however, the user still has great influence on the quality of the final product. A program can do much of the tedious work for you when you are constructing a graph, thus saving much time and effort. Programs allow errors to be corrected and the graph to be redrawn and printed with ease. However, despite the advantages of using a software program, the person producing the graph still needs to be aware of the following:

- Know what type of graph to use in different situations. Sometimes there will be more than one valid choice from a technical viewpoint, but ease of interpretation may be greater with one graph type than another.
- Be aware of the limitations of software programs. The user must oversee the result produced by a program and intervene when necessary. For example, a graph produced with default settings may require some manual adjustments or improvements.

Deciding on the class interval on the axes is also something that you, not the program, should do. Be aware that a software program may produce a graph that displays data correctly but does not, by default, contain some of the essential elements of a graph as described above, such as title, units, or

Box 8.1 Essential information on graphs

Graphs must include some essential information. Throughout this chapter, as the different types of graphs are described, remember that a graph should have these elements:

- Graph title (subtitle or caption may also be included; see p. 15 in Chapter 1 for a description of these terms)
- Lables for each axis
- Scales marked on each axis (with tick marks at all labelled values and possibly at intermediate, unlabelled values)
- Units, where applicable, for each axis (some data being plotted do not have units, for example on an axis with the title "Country")
- Legend (if there are two or more data series)

Some types of graph will not require all of these elements. For example, a pie graph does not have axes. Good practice would be to include all of the relevant elements, by default. Omitting an element should be a conscious decision based on understanding of the type of graph being used. For a graph that shows only one data series, for example, a legend is usually redundant because the graph title or caption will likely provide an adequate description of the data shown.

legend. However, the chapter does not describe in detail the procedure for any particular program. Helpful tutorials can be found on the web (for example, at Microsoft Office Online for creating Excel graphs).

Different Types of Graphs

This chapter describes various forms of graphs and gives some advice about their construction. In a deliberate strategy, all graphs in this book have been created in Microsoft Excel. While other, more powerful, software packages for producing graphs exist, Excel is widely available in most post-secondary institutions, and most students should be able to produce graphs comparable with (or better than!) any of those shown in this chapter. See Table 8.1 for a summary of the major forms of graph discussed, together with the nature and function of each.

Scatter plots

A scatter plot is a graph of point data plotted by (x, y) coordinates (see Figure 8.1 for an example). Scatter plots are usually created to give a visual impression of the direction and strength of a relationship between variables.

Table 8.1 • Types of graphs and their nature or function

Type of Graph	Nature or Function
Scatter plot	Graph of point data plotted by (x, y) coordinates. Usually created to provide visual impression of direction and strength of relationship between variables.
Line graph	Connects values of observed phenomena by lines. Often used to illustrate change over time.
Bar graph	Depicts observed values with horizontal or vertical bars whose *length* is proportional to values represented.
Histogram	Similar to a bar graph but commonly used to depict distribution of a continuous variable. Bar *area* is proportional to value represented. Thus, if class intervals depicted are of different sizes, the column areas will reflect this.
Population pyramid	Form of histogram showing the number or percentage of a population in different age groups of the total population.
Pie graph	Circular graph in which proportions of some total sum (the whole "pie") are depicted as "slices." The area of each "slice" is directly proportional to the size of the data value portrayed.
Graph with logarithmic axes (semi-log graph and log-log graph)	Form of graph with one or both axes having a logarithmic scale. Key intervals on logarithmic axes are exponents of 10. Logarithmic axes allow depiction of data over wide ranges.

Constructing a scatter plot

The *independent* variable (that is, the one that causes change) is depicted on the horizontal *x-axis*, and the *dependent* variable (the variable that changes as a result of change in the independent variable) is plotted on the left vertical *y-axis*. A secondary y-axis (on the right side) can be used to show a second dependent variable with a different scale of measurement. To illustrate the difference between independent and dependent variables, consider the relationship between precipitation and costs associated with flooding. The cost of damage caused by flooding will usually depend on the amount of rainfall: thus, rainfall is the independent variable (*x*-axis) and damage costs are dependent (*y*-axis). Similarly, the severity of injuries associated with motor-vehicle accidents (dependent variable, *y*-axis) tends to increase with vehicle speed (independent variable, *x*-axis).

> ***Be sure to put independent and dependent variables on the correct axes of your graph.***

After points are located on the scatter plot, a "line of best fit" may also be calculated mathematically with regression analysis (see Figure 8.1). A line of best fit may be fitted through the points by eye (that is, your visual impression of the relationship expressed in the form of a line through the points). Note that a line of best fit represents the *trend* of the data—it does not connect

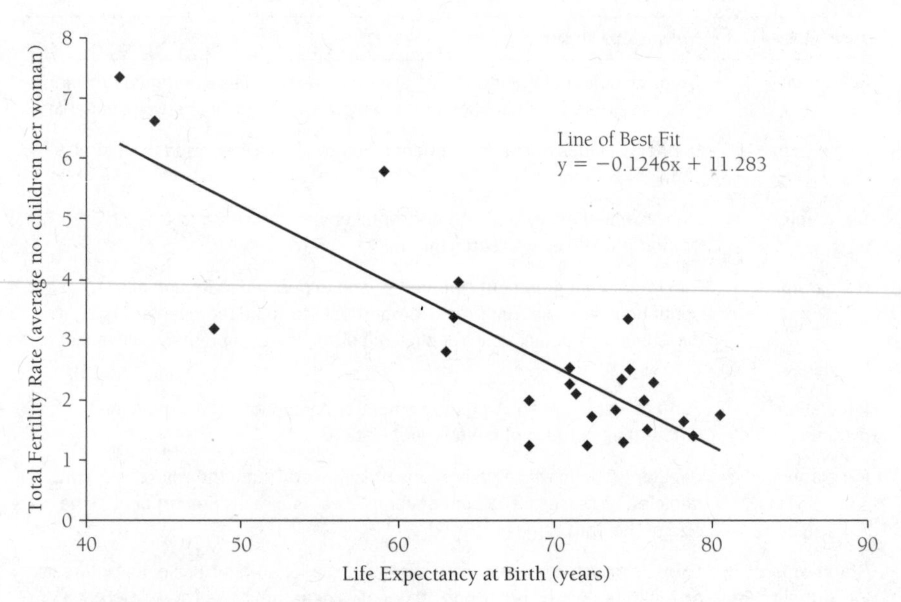

Figure 8.1 ◉ Example of a scatter plot

Total fertility rates vs life expectancy at birth, selected countries, 2005–10

Source: United Nations Population Division (2012).

point to point. It should be drawn as a straight line unless it is clear that the relationship between the variables has a curved pattern.

Line graphs

Typically, line graphs are used to illustrate continuous changes in some phenomenon over time, with any trends being shown by the rise and fall of the line. Line graphs may also show the relationship between two sets of data. Figures 8.2, 8.3, and 8.4 are examples of line graphs.

A line graph should be used only when there are no missing values in the x-axis data. Whereas on a scatter plot those data are actual values, on a line graph they are just labels placed consecutively without regard for the true numerical values, even if in numerical form. That is, if data for 2005 and 2006 are missing, the years 2003, 2004, and 2007 would occupy three consecutive intervals of equal width on a line graph *x*-axis. On a scatter plot, the interval between 2004 and 2007 would be wider than that between 2003 and 2004. Furthermore, while it would likely not make much sense to do so if the variable were years, values on the *x*-axis of a line graph could be placed in the order 2003, 2007, and 2004 because they are just labels, not true numbers. A common mistake is to use a line graph when the data require a scatter plot. For example, a topographic profile with a series of distance (*x*) and elevation (*y*) values requires a scatter plot. The

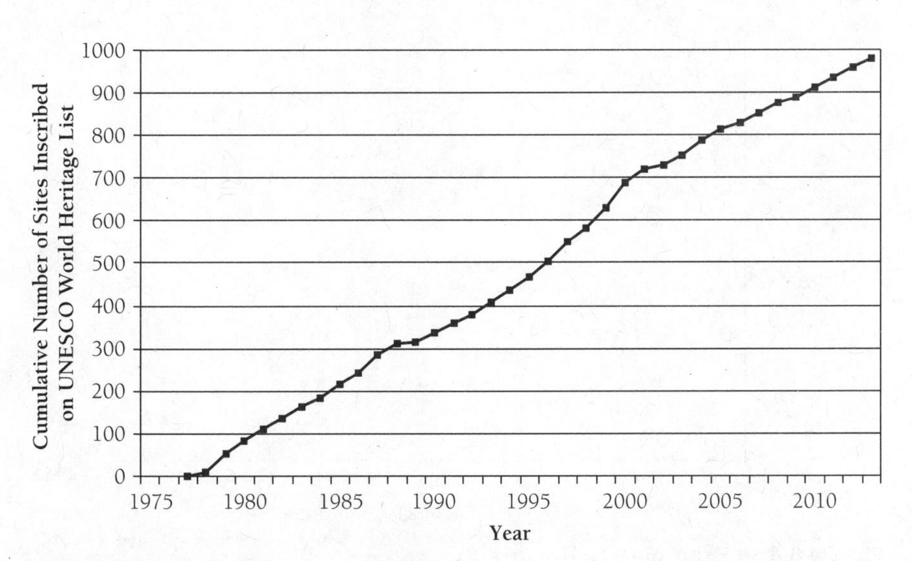

Figure 8.2 ◉ **Example of a line graph**

Cumulative number of sites inscribed on the World Heritage List, 1977–2013

Data Source: World Heritage Centre (2013).

form of the profile will be incorrect if you use a line graph, because the points will not be placed at the correct distance along the *x*-axis.

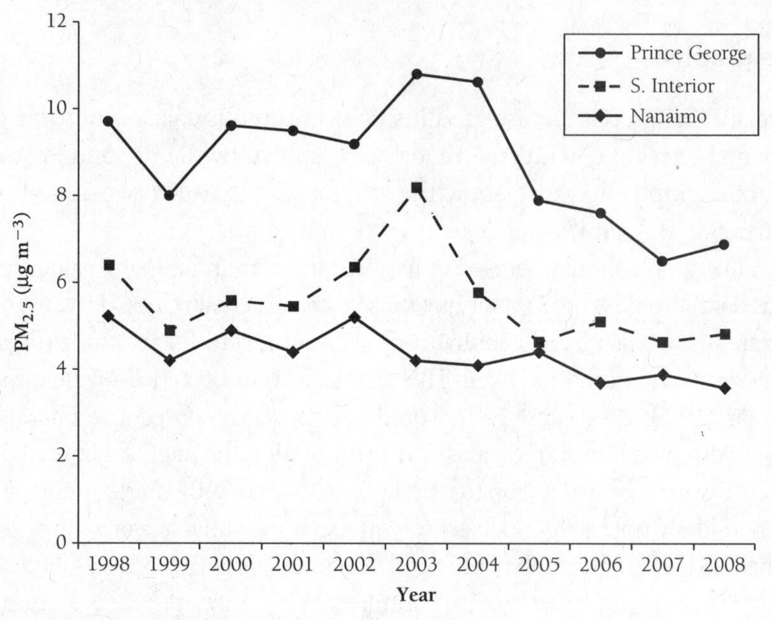

Figure 8.3 ◉ **Example of a line graph**

Trends in fine particulate matter ($PM_{2.5}$) levels in the air at selected locations in British Columbia, 1998–2008

Source: BC Lung Association (2009).

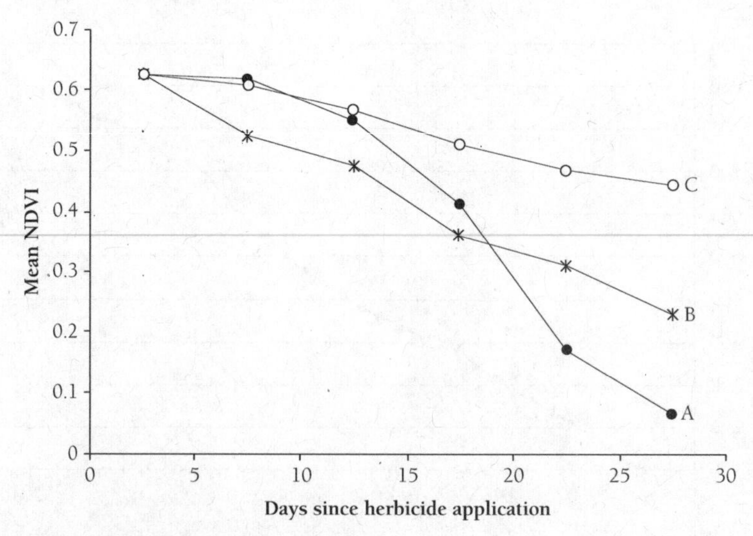

Figure 8.4 ◉ **Example of a line graph**

Mean NDVI (Normalized Difference Vegetation Index)—a measure of vegetation "vigour"—following application of three different herbicides. Figures calculated from NDVI measurements in ten 3 m × 3 m test plots for each herbicide.

Constructing a line graph

Points on a line graph are connected with lines, as the name suggests. Because there are known data values only at the plotted points, lines between points should normally be drawn as straight segments (as shown in Figures 8.2, 8.3, and 8.4). In some cases—for example, where a clear trend is disrupted by a single inconsistent data point (see Figure 8.5)—it is better to draw a smooth curve than to "join the dots" (Pechenik, 2007, p. 179). Make any such decisions very carefully. If a number of lines are shown in one graph, ensure that they can easily be distinguished from one another (see Figures 8.3 and 8.4) by using line patterns, symbols, or colour. If you are showing average (mean) values at each point on your graph, you can usefully provide a visual summary of the variation within the data by, for example, depicting the data range or the standard deviation about the mean (Pechenik, 2007, p. 180). An example is shown in Figure 8.5.

As Pechenik (2007, p. 181) points out:

> Plots of standard deviations or standard errors are always symmetrical about the mean and so convey only partial information about the range of values obtained. If more of your individual values are above the mean than below the mean, the error bars will give a misleading impression about how the

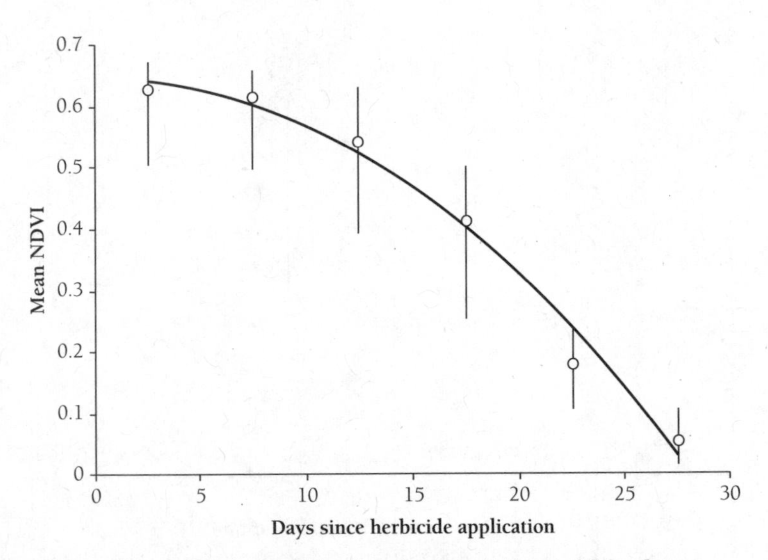

Figure 8.5 ◉ **Example of a graph showing variation within data (mean and range)**

Mean (including trend line) and range of NDVI (Normalized Difference Vegetation Index)—a measure of vegetation "vigour"—measurements in ten 3 m x 3 m test plots following the application of Herbicide A.

> data are actually distributed. If your graph is fairly simple, you may be able to achieve the best of both worlds, indicating both the range and standard deviation (or standard error).

An example is provided in Figure 8.6.

If you do include indicators of variation in your graph, make sure that you include, in your caption notes to the figure, details of what you have plotted, together with the number of measurements associated with each mean (Pechenik, 2007, p. 182).

Line graphs (and scatter plots) will sometimes display variables that have different scales of measurement. This can be done by using vertical axes on the left and right sides of the graph to depict the different scales. A climograph, portraying mean monthly temperature and mean monthly precipitation for a place, is a common graph with two axes. (On a climograph, temperature is displayed as a line graph and precipitation is displayed as a bar graph—the type of graph that is discussed next.) Figure 8.7 illustrates the use of two vertical axes on one graph. Sulphate and calcium are measured as concentrations on a scale of milliequivalent per litre, while pH measures the acidity or alkalinity of a substance on a scale that ranges from 0 to 14.

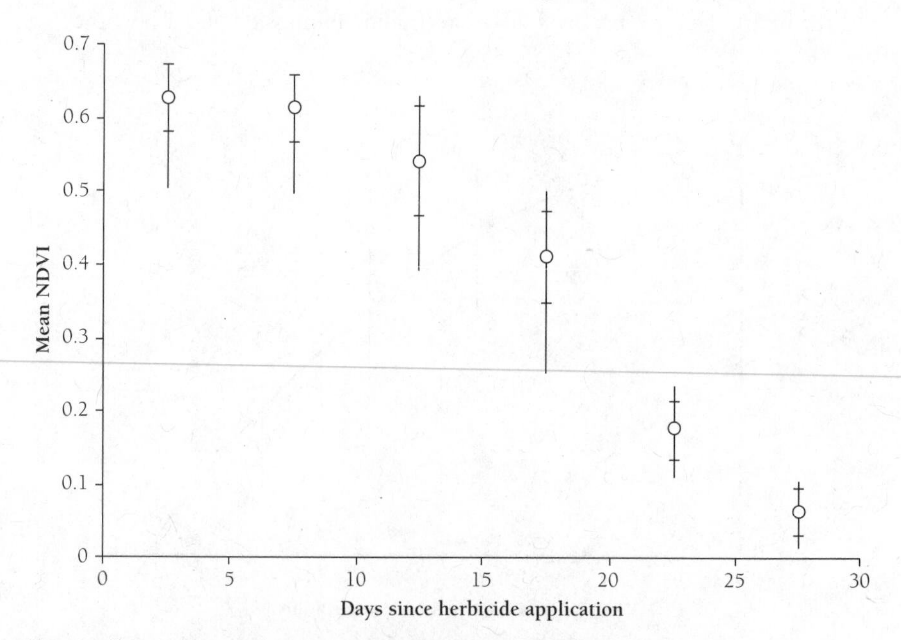

Figure 8.6 ◉ Example of a graph showing variation within data (data range, mean, and standard deviation)

Variations about the mean (range and standard deviation) in NDVI (Normalized Difference Vegetation Index)—a measure of vegetation "vigour"—measurements in ten 3 m x 3 m test plots following the application of Herbicide A.

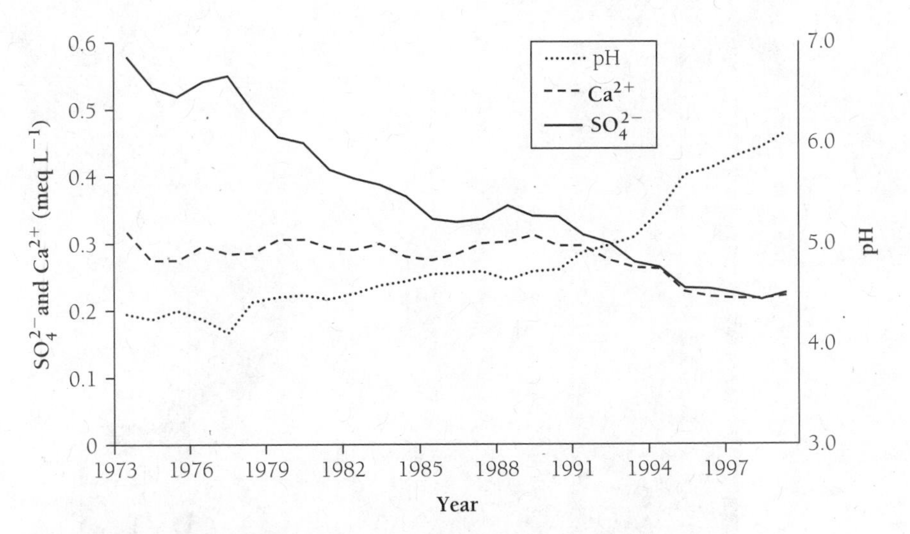

Figure 8.7 ◉ **Example of a line graph with two vertical axes**

Changes in lake chemistry, Clearwater Lake, Ontario, 1973–1999. SO_4^{2-} is sulphate, Ca^{2+} is calcium, and pH measures the acidity or alkalinity of a substance.

Source: Jeffries et al. (2003).

Bar graphs

Bar graphs are of two main types: *vertical* and *horizontal*. Nicol and Pexman (2003, p. 27) point out that vertical bar graphs are normally preferred since it is conventional to show the independent variable on the horizontal axis. With a horizontal bar graph, the independent variable appears on the vertical axis. Figures 8.8 and 8.9 show the two types.

In bar graphs the *length* of each bar is proportional to the value it represents. It is in this regard that bar graphs differ from histograms, with which they are sometimes confused. Histograms use bars whose *areas* are proportional to the value depicted.

Bar graphs are a commonly used and easily understood way of taking a snapshot of variables at one point in time, depicting data in groups, and showing the size of each group. Figure 8.10 achieves all of these ends in a single graph.

Bar graphs can also be used to show the components of data as well as data totals. See Figure 8.11 for an example of such a *subdivided bar graph*. It is possible to go one step further and represent data in the form of a *subdivided 100 per cent bar graph* (see Figure 8.12 for an example).

As Figure 8.13 illustrates, bar graphs can be used to portray negative as well as positive quantities. In this example, a horizontal graph was chosen instead of a vertical graph because it improved the display of data labels on the

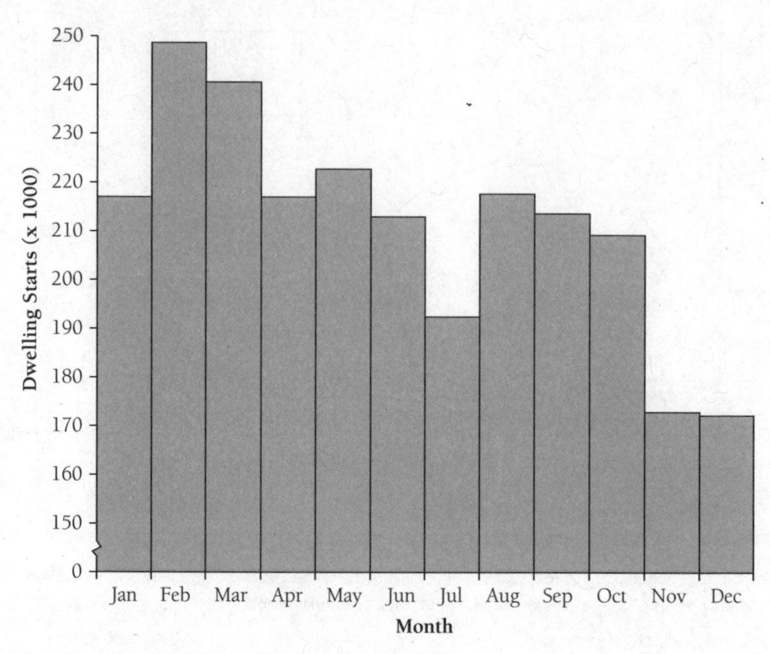

Figure 8.8 ◉ **Example of a vertical bar graph**

Dwelling starts in Canada, seasonally adjusted data at annual rates, 2008

Source: Canada Mortgage and Housing Corporation (2009).

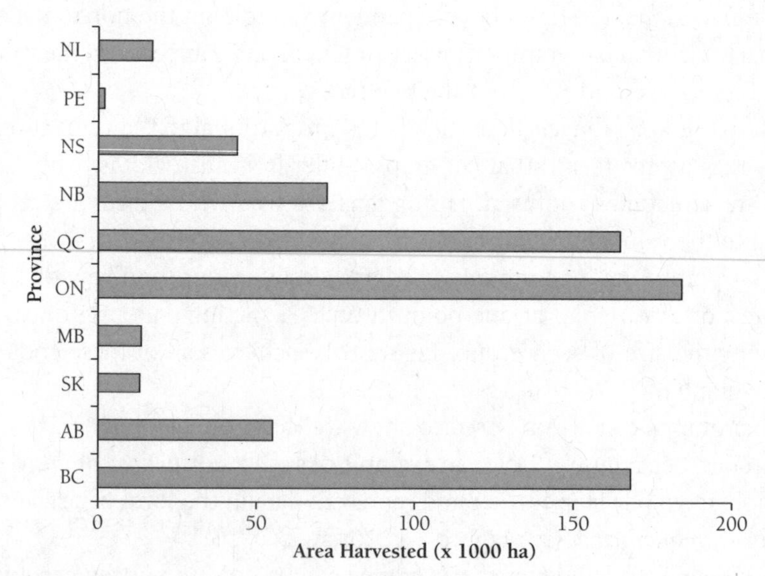

Figure 8.9 ◉ **Example of a horizontal bar graph**

Forest area harvested, by province, 2007

Source: Natural Resources Canada.

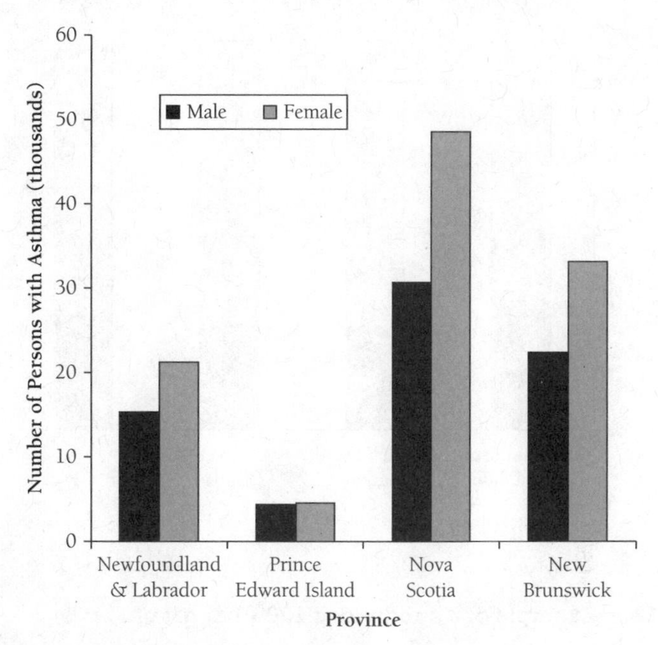

Figure 8.10 ◉ **Example of a bar graph**

Number of persons with asthma in Atlantic provinces, 2008

Source: Statistics Canada (2009).

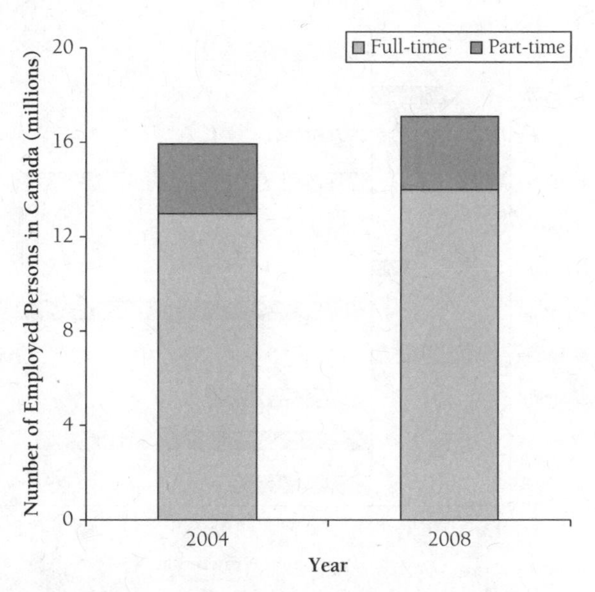

Figure 8.11 ◉ **Example of a subdivided bar graph**

Comparison of part-time and full-time employment in Canada, 2004 and 2008

Source: Statistics Canada (2009).

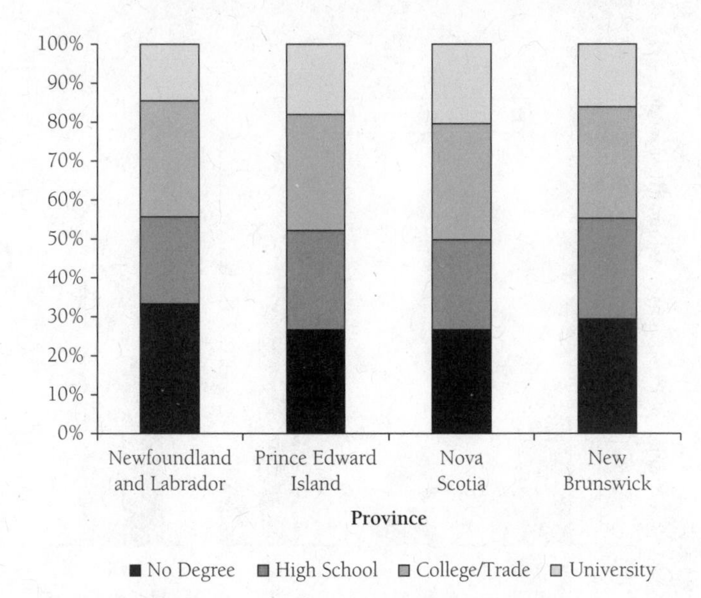

Figure 8.12 ◉ **Example of a subdivided 100% bar graph**

Highest level of education attained in Atlantic provinces, 2006 Census

Source: Statistics Canada (2009).

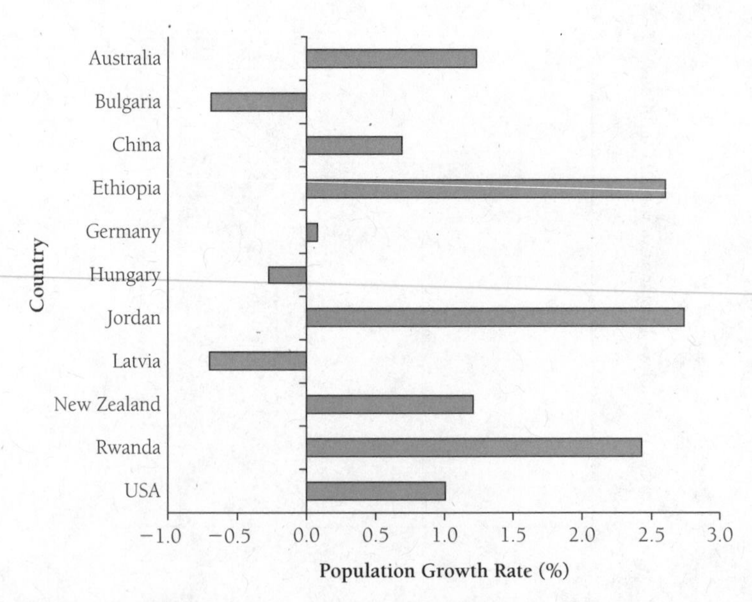

Figure 8.13 ◉ **Example of a bar graph depicting positive and negative values**

Annual population growth rates for selected countries, 2000–5

Source: United Nations Population Division (2012)..

vertical (independent variable) axis. More efficient use of space in displaying country names of different length resulted in a more pleasing graph than if the countries had been shown on the horizontal axis.

Constructing a bar graph

Bar graph scales normally commence at zero, although this is not critical. If a scale does start at zero, make sure zero is labelled, since it is as valid a number as any non-zero value. A common error is to begin labelling at the first non-zero number on the scale, leaving a blank space where the numeral zero should appear.

In Figure 8.8, the vertical axis contains a break between 0 and 150 000. This was done to highlight differences between data values by cutting off the first 150 000 dwelling starts that occurred in all months. Showing the full range of the scale from 0 to 250 000 would have created bars that were less distinct visually. Note that using the axis break may give the reader a false impression of the actual difference between lower and higher values, so this technique should be used carefully.

When the graph is being designed, it is important to consider the sequence of items being depicted. In general, the items should be listed in order of importance to the viewer. However, for simple comparisons in a horizontal bar graph, it is best to arrange the bars in ascending order of length from bottom to top. That said, you must also be aware that some data sets are listed, by convention, in particular orders. For example, in Canada, bar graphs showing data by province and territory often proceed from one side of the country to the other in geographical order (British Columbia to Newfoundland and Labrador, or vice versa), with the territories placed at the far right side. Bar graphs should follow such conventions. The bars should be separated from one another, reflecting the discrete nature of the observed values (Jennings, 1990, p. 18), and the space between the bars should be about one-half to three-quarters of their width. However, when there is continuity between adjacent categories on the *x*-axis (such as continuity in time on a bar graph showing data for consecutive months or years), it is permissible to leave no space between the observations, as in Figure 8.8.

> ***Make sure your graph is labelled fully and correctly.***

Finally, add the title, axis labels, legend, and references.

Histograms

Histograms are mainly used to show the frequency distribution of values of a continuous variable. Continuous data, such as plant height, rainfall, or temperature, could have any conceivable value within an observed range, including fractional or decimal values. In discrete data, however, such as number of plants or animals, there are no decimals or fractional numbers. For examples of histograms, see Figure 8.14. This figure shows two histograms based on the same data set but using different class intervals. Calculation of class intervals is explained on page 176.

Histograms may be confused with vertical bar graphs or column graphs, but there is a technical difference. Strictly speaking, histograms depict frequency through the area of the column, whereas in a column graph frequency is measured by column *height* (Australian Bureau of Statistics, 1994, p. 120). Thus, while histograms usually have bars of equal width, if the class intervals are of different sizes the columns should reflect this. For example, if one class interval on a graph were 0°C to 5°C and a second were 5°C to 15°C, the second should be drawn twice as wide as the first. Because a histogram with inconsistent class intervals requires a higher level of understanding and interpretation by the reader, consistent class intervals should be used whenever possible. Remember that the purpose of using a graph is to make numerical information easier for a person to understand. Unequal class intervals may actually mislead a reader who looks at the graph quickly and does not consider the individual class sizes carefully. The phenomenon whose size is being depicted is plotted on the horizontal *x*-axis. Frequency of occurrence is plotted on the vertical *y*-axis. The frequency is the number of occurrences of the measured variable within a specific class interval, such as number of days with mean temperature in a given range.

Example of data set

The data set used to construct the histograms in Figure 8.14 will not be shown in full detail here, but its characteristics are as follows:

- Data for the Yarmouth, Nova Scotia, were obtained from the Environment Canada (2013) website.
- Observations are mean daily temperatures, measured in degrees Celsius, for the year 2007. The data set therefore contains 365 values.
- Data are provided with precision of one decimal place.
- Minimum and maximum values are –12.6°C and 21.4°C, respectively. The mean annual temperature is 6.9°C.

Constructing a histogram

As Figure 8.14 illustrates, the method you choose to construct your histogram can have a significant effect on the appearance of the graph you finally produce. Figures 8.14(a) and (b) are based the same data set (see above).

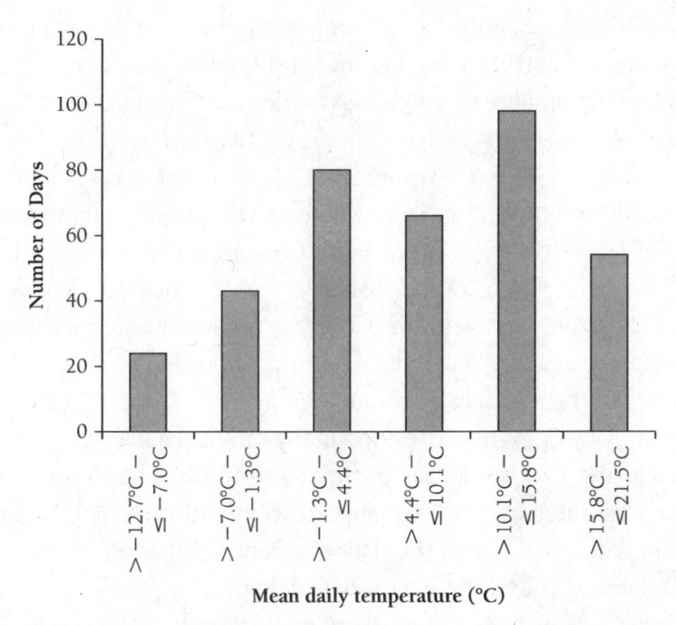

(a) Distribution of mean daily temperature, Yarmouth, Nova Scotia, 2007

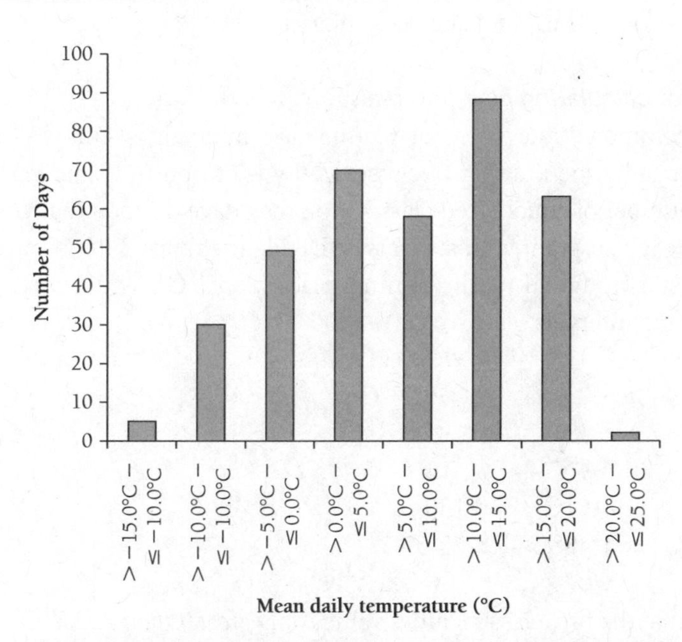

(b) Distribution of mean daily temperature, Yarmouth, Nova Scotia, 2007

Figure 8.14 ◉ **Examples of histograms based on the same data but different class intervals**

Source: Environment Canada (2013).

However, each was created by a different method of calculating class intervals and frequency distributions. Figure 8.14(a) splits the data range evenly on the basis of the *number* of *x*-axis classes desired (in this case six). Figure 8.14(b) shows the data on the basis of the desired *size* of the *x*-axis classes (in this case 5 C°). (Note that *differences* in temperature, or a temperature *range*, are described with units of Celsius degrees, not degrees Celsius, which are the units of a given temperature.) There is no definite rule governing the number of classes in a frequency distribution. Choose too few, and information could be lost through a large summarizing effect. That is, the picture will be too general. With a lot of classes, too many minor details may be retained, thereby obscuring major features of the distribution. The two methods of working out class intervals and frequency distributions presented here require that you calculate the range of the data set. Range is the difference between the highest data value and the lowest data value.

Range is the difference between the highest data value and the lowest.

The highest mean daily temperature in Yarmouth in 2007 was 21.4°C; the lowest was –12.6°C. Therefore, the range is:

$$21.4 - (-12.6) = 34.0\ \text{C}°.$$

The next step is to calculate class intervals.

Methods of calculating class intervals

1. One common strategy for calculating class intervals is simply to divide the range by the number of classes you wish to portray. The result will be a number of equal-sized classes. You may have decided that you wish to have a histogram with six classes. Divide the range 34.0 C° by 6, and the result (rounded to one decimal place) is 5.7 C°. Thus, we have the following intervals:

   Class 1 > -12.7°C — ≤ -7.0°C
   Class 2 > -7.0°C — ≤ -1.3°C
   Class 3 > -1.3°C — ≤ 4.4°C
   Class 4 > 4.4°C — ≤ 10.1°C
   Class 5 > 10.1°C — ≤ 15.8°C
   Class 6 > 15.8°C — ≤ 21.5°C

   Note that the first class includes values only *greater than* –12.7°C and *less than or equal to* –7.0°C, a range of 5.7 C°. The actual minimum value in the data set, with one decimal place, is –12.6°C, but with a precision of one decimal place there can be no values between –12.6°C and –12.7°C. A starting point of –12.7°C was chosen to ensure that the range of the first class appears as 5.7 C° like all the other classes. An alternative would be to show the class intervals as "–12.6°C to –7.0°C," "–6.9 to –1.3°C," and so on, since values between –6.9°C and –7.0°C cannot be stored with only one decimal place. However, the class interval range in that format might then be misunderstood as being 5.6 C° instead of 5.7 C°. The 0.1 C° that is apparently "missing" between the upper limit of one class and the lower limit of the next class corresponds to the precision of the data and is part of the class interval range (e.g., 5.6 C° + 0.1 C° = 5.7 C°).

Confusion can occur at class interval boundaries if the labelling is not done carefully. Any given value can belong to only one class. In this example, use of the *greater than* and *less than or equal to* symbols ensures not only that the class interval range is clear, but also that there is also no ambiguity about which class a value belongs to: for example, –7.0°C belongs to Class 1. However, if the class boundaries were shown as "Class 1: –12.7°C to –7.0°C," "Class 2: –7.0°C to –1.3°C," and so on, in which class would you put a value of –7.0°C? The same situation occurs regardless of the units of measurement being used (for example, $1, 1 cm, 1 m, 0.01 gram, 1 tonne) or the precision of the data.

If discrete class intervals are used, corresponding discrete values can be generated from non-discrete data by rounding off or truncating. Consider a data set for which you want to show class intervals in whole dollars. The ranges might be $0–$39, $40–$79, and so on. Into which class would you put a value of $39.77? Rounding off to the nearest whole number places $39.77 in the $40–$79 class. Truncating $39.77 to $39 places it in the $0–$39 class. More important than which method you choose is to apply it *consistently* as you work through the data set. You should avoid alternating between rounding off and truncating.

2. An alternative, but closely related, strategy is to choose, rather than calculate, the class interval size and a starting point for the classes. Would it be useful to start or end the class intervals at some points other than those fixed by the high and low points of the data set? The intervals shown in Table 8.2 were chosen for the data set described above.
3. Class intervals that extended in value beyond the upper and lower limits of the data range were selected. The first class begins at –18°C, and the class interval is 6 C°. For this example, the resulting class boundary between Classes 3 and 4 is useful to portray because 0°C is an important physical threshold, the freezing point of water. Other data sets may have important values that should correspond with a class boundary. Some people have a preference for creating class intervals with numbers divisible by 5 or 10, such as 5, 10, or 50, rather than numbers like 5.7, 6, or

Table 8.2 ◉ **Classes for Yarmouth mean daily temperatures, 2007, with 6 C° intervals starting at –18°C**

Class 1	> −18.0°C – ≤ −12.0°C
Class 2	> −12.0°C – ≤ −6.0°C
Class 3	> −6.0°C – ≤ 0.0°C
Class 4	> 0.0°C – ≤ 6.0°C
Class 5	> 6.0°C – ≤ 12.0°C
Class 6	> 12.0°C – ≤ 18.0°C
Class 7	> 18.0°C – ≤ 24.0°C

64, although it is not a strict rule. With this strategy, intervals of 5 C°, starting at –5°C, the classes produced are shown in Table 8.3 and the resulting graph is shown in Figure 8.14b.

The two methods discussed above are both examples of creating classes with *equal intervals*. A number of other methods are used, but the details are beyond the scope of this book. If the data set has a very large range between high- and low-frequency values or contains a small number of values widely separated from the bulk of the data, a more sophisticated method than equal intervals might be used to portray the data more effectively. Slocum et al. (2009) describe six methods for choosing class intervals (equal intervals, quantiles, mean-standard deviation, maximum breaks, natural breaks, and optimal), and they discuss the advantages of each method for data sets with particular characteristics. Once you have worked out class intervals, the next step in the construction of a histogram is to create a frequency table that will allow you to work out the total number of individual items of data that will occur in a particular class.

Table 8.3 ◉ **Classes for Yarmouth mean daily temperatures, 2007, with 5 C° intervals starting at –15°C**

Class 1	> −15.0°C – ≤ −10.0°C
Class 2	> −10.0°C – ≤ −5.0°C
Class 3	> −5.0°C – ≤ 0.0°C
Class 4	> 0.0°C– ≤ 5.0°C
Class 5	> 5.0°C – ≤ 10.0°C
Class 6	> 10.0°C – ≤ 15.0°C
Class 7	> 15.0°C – ≤ 20.0°C
Class 8	> 20.0°C – ≤ 25.0°C

Constructing a frequency table

As shown in Table 8.4, constructing a frequency table is simply a matter of going through the set of data, placing a tally mark against the class into which each data value falls, then adding up the tally marks to find the frequency with which values occur in each class. Microsoft Excel has a tool for creating a frequency table from a list of values that is especially useful for large sets of data. Table 8.4 uses the 5 C° class intervals described above.

The observed frequency is then plotted on the y-axis of the histogram, and the classes are plotted on the *x*-axis (see Figure 8.14b).

Population pyramids or age–sex pyramids

Population pyramids are a special form of histogram used to show the number or, more commonly, the percentage of a total population in different age groups. They also illustrate the male–female composition of that population. Figure 8.15 is an example of a population pyramid.

Constructing a population pyramid

Although a population pyramid is a form of histogram, a few peculiarities do bear noting. As Figure 8.15 illustrates, a population pyramid is drawn on one vertical axis and two horizontal axes. The vertical axis represents age and is usually subdivided into five-year age cohorts (for example, 0–4, 5–9, 10–14, 15–19 years). However the size of those cohorts may be changed (for example, to 0–9, 10–19, or 0–14, 15–19) depending on the nature of the raw data and the purpose of your pyramid. Remember from the discussion

Table 8.4 ◉ **Frequency table of data for Yarmouth mean daily temperatures, 2007**

Class	Classes (°C)	Tally of occurrence	Frequency
1	> −15.0°C −≤ −10.0°C	𝍸	5
2	> −10.0°C −≤ −5.0°C	𝍸 𝍸 𝍸 𝍸 𝍸 𝍸	30
3	> −5.0°C −≤ 0.0°C	𝍸 𝍸 𝍸 𝍸 𝍸 𝍸 𝍸 𝍸 𝍸 IIII	49
4	> 0.0°C −≤ 5.0°C	𝍸 𝍸 𝍸 𝍸 𝍸 𝍸 𝍸 𝍸 𝍸 𝍸 𝍸 𝍸 𝍸 𝍸	70
5	> 5.0°C −≤ 10.0°C	𝍸 𝍸 𝍸 𝍸 𝍸 𝍸 𝍸 𝍸 𝍸 𝍸 𝍸 III	58
6	> 10.0°C −≤ 15.0°C	𝍸 𝍸 𝍸 𝍸 𝍸 𝍸 𝍸 𝍸 𝍸 𝍸 𝍸 𝍸 𝍸 𝍸 𝍸 𝍸 𝍸 III	88
7	> 15.0°C −≤ 20.0°C	𝍸 𝍸 𝍸 𝍸 𝍸 𝍸 𝍸 𝍸 𝍸 𝍸 𝍸 𝍸 III	63
8	> 20.0°C −≤ 25.0°C	II	2

of histograms, however, that if you depict different-sized cohorts in the same pyramid, the area of each bar must reflect that variability. For example, a 0–14 cohort would be half as wide as a 15–44 cohort on the same graph. On either side of the central vertical axis are the two horizontal axes. The axes on the left of the pyramid shows the percentage (or number) of males while the one on the right shows the figures for females. You will also see from Figure 8.15 that the zero point for each of the horizontal axes is in the centre of the graph. As a final note, it may also be helpful for your reader if you include a statement of the total population depicted. It is possible to create a population pyramid in Microsoft Excel, but it is not as straight-forward as producing many other types of graph. The trick is that one-half of the population (usually males, portrayed on the left side) must be entered as *negative* values. Then an option for the horizontal axis is selected to *display* values on both sides of zero as positive numbers.

By convention, a population pyramid shows males on the left and females on the right.

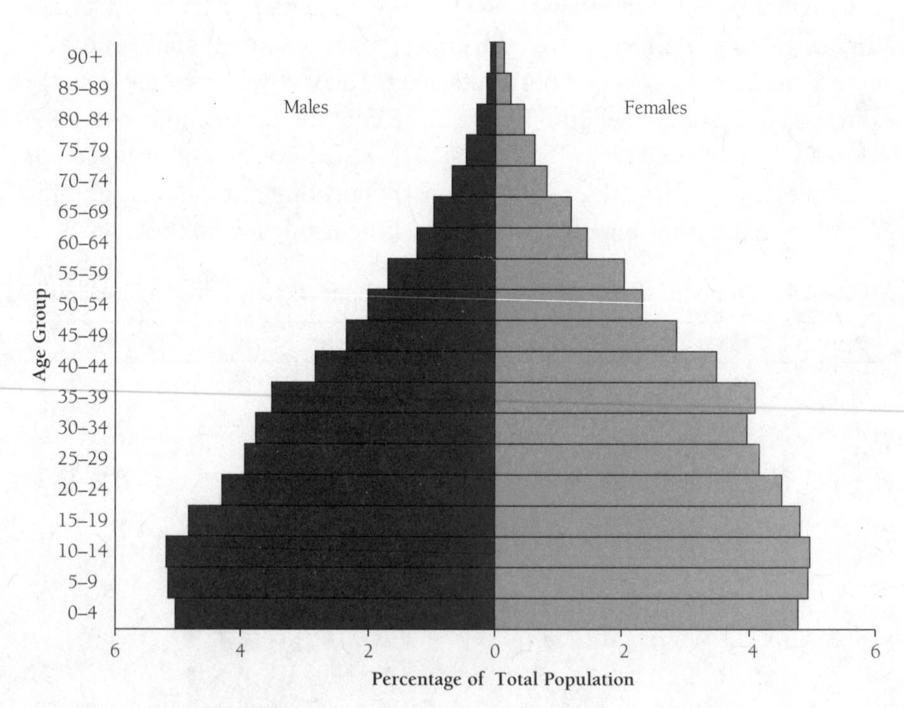

Figure 8.15 ● Example of a population pyramid

Age–sex structure of Mexico's population, 2010 (total population = 117 886 000)

Source: United Nations Population Division (2012).

Pie graphs

Pie graphs show how a whole is divided up into parts and what share or percentage belongs to each part. Pie graphs are a dramatic way of illustrating the relative sizes of portions of some complete entity. For example, a pie graph might show how a budget is divided up or who receives what share of some total. See Figure 8.16 for an example.

Constructing a pie graph

It is necessary to match the 360 degrees that make up the circumference of a circle with the percentage size of each of the variables to be graphed. Simply, this is achieved by multiplying the percentage size of each variable by 3.6 to find the number of degrees to which it equates. Obviously, if the values have not already been translated into percentages of the whole, this will need to be done first. It is best not to have too many categories (or "slices") in a pie graph since this creates visual confusion. Generally, no segment should be smaller than about 3 per cent of the total. This may require that the data set be rearranged

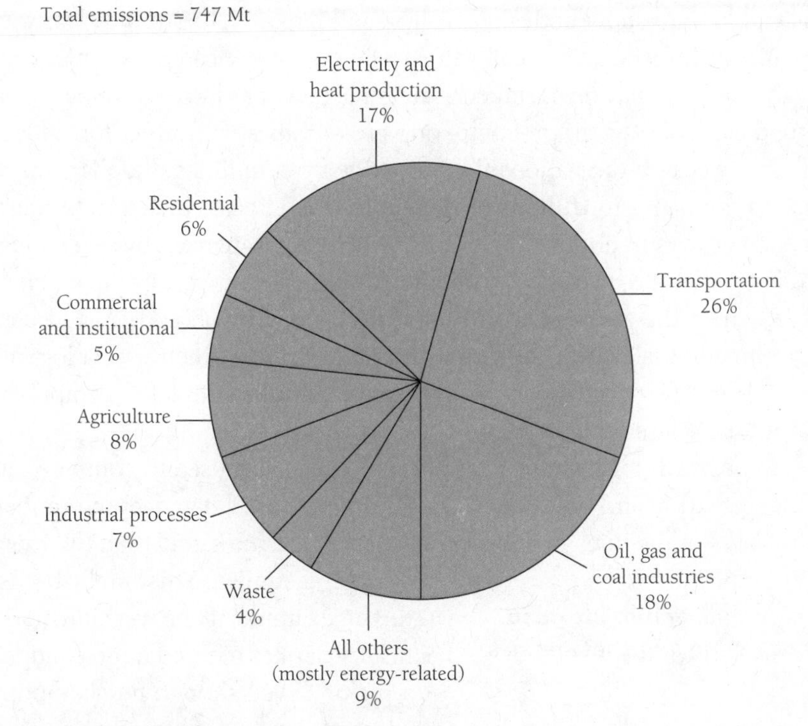

Figure 8.16 ◉ Example of a pie graph

Greenhouse gas emissions by activity sector, Canada, 2005

Source: Statistics Canada (2009).

so that some classes are grouped together. Pie graphs should also inform the reader of the *total value* of categories plotted, as shown, for example, by the statement in Figure 8.16 of the total quantity of greenhouse gas emissions (747 Mt). There is little point in telling your readers percentages without letting them know exactly how much that percentage represents in absolute terms. The absolute value of each category may be shown along with its percentage value, but keep the overall appearance in mind and be sure not to create a graph that is too cluttered with labels.

A pie graph should always make clear the total value of all categories it illustrates.

Graphs with logarithmic axes

Strictly speaking, a graph with one or more logarithmic axes is not a new type of graph; rather, the basic form of the graph is commonly a scatter plot or line graph. Logarithmic axes are used primarily when the range of data values to be plotted is too great to depict on a graph with arithmetic axes. If the range of data to be graphed covers several magnitudes, a logarithmic axis will help to show both the smallest and largest values more clearly. On a regular arithmetic axis, small values may not be noticed in comparison to large values and may be mistaken as missing data or as having a value of zero. A good example is a graph comparing gross domestic product for different countries because the national figures range from millions of dollars for the smallest countries to trillions of dollars for the largest countries. Similarly, showing growth in population figures, which might increase by several magnitudes, sometimes necessitates the use of a logarithmic axis. Compare Figure 8.17a, where the graph of world population growth was constructed using an arithmetic scale, with the graph in Figure 8.17b, which uses a logarithmic scale on the population axis. Figure 8.18 is an example of a graph with logarithmic scales on both axes.

Graphs with logarithmic axes are sometimes also used to compare rates of change within and between data sets. Despite vast differences in numbers, if the *slopes* of the lines in a logarithmic graph are the same, then the *rate* of change is similar. This might be useful, for example, if one were illustrating historical rates of population change in a region. Even though *absolute* population growth might be much greater in one period compared to another, the *rates* of population growth in the two periods might be similar. Notice in Figure 8.17b, for example, that the rate of global population growth from

Logarithmic graphs are good for depicting large data ranges.

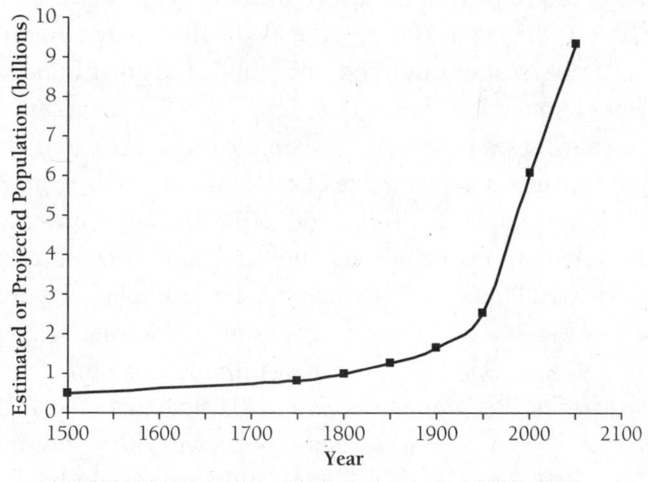

(a) World population, estimated or projected, 1500–2050

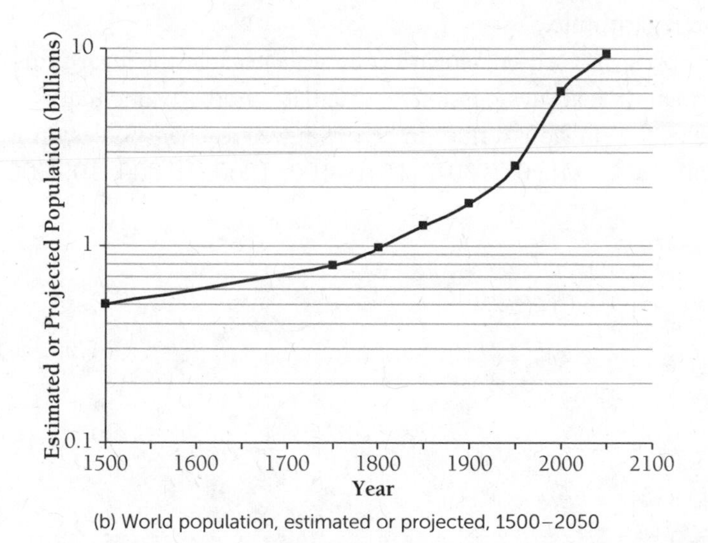

(b) World population, estimated or projected, 1500–2050

Figure 8.17 ◉ Comparison of data displayed on (a) arithmetic vertical axis and (b) logarithmic vertical axis

Source: U.S. Census Bureau (2009).

2000 to 2050 is predicted to be similar to the period from 1900 to 1950. The graph shows that the rate was greater during the 1950–2000 period than during the preceding or succeeding periods because the line is steeper.

Before discussing this form of graph any further, it might be helpful to say a little about logarithms. The logarithm of a number (in base 10) is the power to which 10 must be raised to give that number. For example, the log of 100 is 2 because $10^2 = 100$ (that is, 10 raised to the power 2 = 100). Thus, the log of 10 is 1; the log of 100 is 2; the log of 1000 is 3 and so on.

Second, simple scatter plots and line graphs use an arithmetic scale on their axes (for example, 1, 2, 3, 4 or 0, 2, 4, 6 . . .), where a constant numerical difference is shown by an equal interval on the graph axes. In contrast, on a logarithmic scale the numerical value of each interval on the graph increases *exponentially* (for example, 10, 100, 1000, 10 000) and the lines within each cycle (each cycle is an exponent of 10) of the graph become progressively closer together (see Figures 8.17b and 8.18 for examples). Figure 8.17b is an example of a *semi-log* graph. It has a logarithmic *y*-axis and an arithmetic *x*-axis. (Histograms and bar graphs can also be drawn with a semi-logarithmic scale if the variable to be depicted on the *y*-axis has a particularly large range.) Figure 8.18 is a *log-log* graph—both axes are logarithmic.

> ***Semi-log graphs have one logarithmic axis; log-log graphs have two.***

Zero is never used on a logarithmic scale because the logarithm of zero is not defined. It is always possible to calculate another value that is one-tenth as large as the previous value, so zero is never reached. Scales on log paper start with . . . 0.001, 0.01, 0.1, 1, 10, 100, 1000, 10 000, 100 000 . . . (or

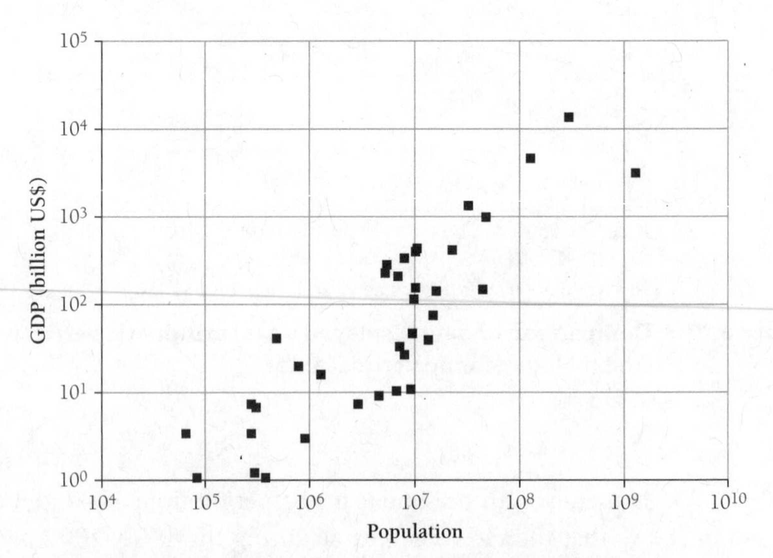

Figure 8.18 ◉ Example of a graph with two logarithmic axes (a "log-log" graph)

Relationship between GDP and population, selected countries, 2008

Source: U.S. Department of Agriculture, Economic Research Service (2009).

any other exponent of ten). If you look at Figure 8.18 as an example, you will see that the *x*-axis begins at 1, whereas the *y*-axis begins at 10 000. The determination about the value with which to start the axis is made on the basis of the smallest data point. For example, if the smallest value being plotted were 35 000, you would start the axis at 10 000, not 1000 or 100 000.

Zero never appears on a logarithmic axis.

Constructing a graph with a logarithmic scale

Consider the maximum and minimum values of the data sets to be plotted. In looking, for example, at the world population over the period 1500–2050 (Figure 8.17), we see that the population grew from 500 million to 9.3 billion people. To accommodate this range of data, the graph needed to have two logarithmic cycles, with the first starting at 100 million (or 0.1 billion), and the second at one billion. If world population exceeded 10 billion, a third logarithmic cycle would have to be included. Because the years are plotted arithmetically, the graph was drawn in semi-log format. The same procedures apply for log-log graphs (for example, see Figure 8.18) except, of course, that it is necessary to consider the number of cycles for both axes, not just one.

Tables

Tables present related facts or observations in an orderly, unified manner. They are used most commonly for summarizing information more clearly than in long passages of text. For example, consider how much text would be required to present the information in Table 8.5—and how boring it would be to read that text! Using a table means that you are not writing as many words to present the information, but if the table is designed well the reader will still obtain information. Tables can be effective for organizing and communicating large amounts of information, especially numerical data, although you should not make the mistake of trying to convey too much information at once. The main reasons for using tables are these:

- To facilitate comparisons;
- To reveal relationships;
- To save space.

Tables should be self-explanatory. This requires an informative title and good headings for the columns and

A good table is comprehensible by itself.

Table 8.5 ◉ Example of a table
Land and water areas of Canadian provinces and territories

	Area (km²)			
Province or Territory	**Land**[a]	**Water**	**Total**	**Percentage of Canadian Total**[b]
Newfoundland & Labrador	373 872	31 340	405 212	4.06
Prince Edward Island	5660	n/a[c]	5660	0.06
Nova Scotia	53 338	1946	55 284	0.55
New Brunswick	71 450	1458	72 908	0.73
Quebec	1 365 128	176 928	1 542 056	15.44
Ontario	917 741	158 654	1 076 395	10.78
Manitoba	553 556	94 241	647 797	6.49
Saskatchewan	591 670	59 366	651 036	6.52
Alberta	642 317	19 531	661 848	6.63
British Columbia	925 186	19 549	944 735	9.46
Yukon Territory	474 391	8 052	482 443	4.83
Northwest Territories	1 183 085	163 021	1 346 106	13.48
Nunavut	1 936 113	157 077	2 093 190	20.96
Canada	9 093 507	891 163	9 984 670	100.00

Notes:
a. This note was inserted to illustrate how a table footnote might look. See the discussion below to find out about the purpose of such notes.
b. "Percentage of Canadian total" is calculated from the combined totals of land and water in each province or territory.
c. n/a stands for "Data not available."

Source: Canadian Council of Forest Ministers, National Forestry Database (http://nfdp.ccfm.org). Reproduced with permission of Natural Resources Canada. 2014.

labels for the rows. If a title alone is not comprehensive enough, a subtitle can also be included. Important trends or notable values in the table may be referred to and discussed in accompanying text, but data should not be repeated extensively. Table 8.5 is an example of a well laid-out table.

Elements of a table

The main elements of a table are as follows:

1. *Table number*. Each table should have a unique number (for example, Table 1) allowing it to be referred to in the text.
2. *Table title (and possibly a subtitle)*. The title, which is placed one line above the table itself, should be brief and allow any reader to fully comprehend

the information presented without reference to other text. The title should answer "what," "where," and "when" questions. If a title alone would have to be too long in order to convey necessary information or clarification, use a subtitle as well.

3. *Column headings*. Headings are necessary to explain the meaning of data appearing in the columns. Specify the units of measurement (for example, $, mm, litres) within the column headings and do not repeat the units for each value in the table. Since only a small amount of space is available for headings, they must be concise but not so concise that they become ambiguous. The column headings are usually separated from the data by a horizontal line.
4. *Row labels*. Names of the entities or the categories for which data are shown in the rows.
5. *Stub*. The first (left-hand) column of a table. Consists of a vertical list of categories. In the sample column above, the stub consists of the provinces and territories.
6. *Table notes*. Notes appear below the table as footnotes and provide supplementary information to the reader, such as limitations that apply to some of the reported data. Table notes may also be used for explanation of any unusual abbreviations or symbols or to provide more detailed information than there is space for in the table itself. Notes are often "numbered" with letters rather than numbers to avoid confusion with the numbers displayed in the table. If numbers are used, they should start at 1 for each table.
7. *Table source*. An indication of the source from which the data were derived or the place from which the table was reproduced should be provided. An accepted form of referencing must be used. See Chapter 6 for more information on referencing.

In the end, always be sure that the way you present a table or figure helps your reader understand it completely and correctly. If you have any doubts, present the figure or table to friends as a stand-alone document and ask them if it makes sense!

Designing a table

A table must be designed well in order to present the data clearly and efficiently. The aspects to pay attention to are as follows:

1. *Use of dividing lines*: Lines should be used where it is useful or necessary to provide the reader with visual cues indicating material that has something in common. Extraneous lines increase visual clutter without adding

value. In scholarly journals, minimal use of lines is encouraged in order to present a clean or tidy appearance. For example, no lines were placed between the provinces and territories in Table 8.5. However, a horizontal line was placed between the rows for Nunavut and Canada because Nunavut is a similar entity to the other provinces and territories, but a different entity from Canada. Line styles can vary (for example, thicker lines or double lines can be used for more important separations). Use lines only where they are necessary to emphasize separation of material. Poorly placed lines can confuse the reader, especially if the placement divides the table illogically. Putting some effort into placement and/or types of lines can markedly improve the appearance and clarity of a table. Avoid placing lines of similar appearance around all cells in a table. However, the bottom of the table should be marked with a horizontal line to separate it visually from the subsequent text.

2. *Type styles*: Boldface or italics can be used to draw attention to similarities or differences in different parts of the table.
3. *Column width*: Columns should be adjusted to a width that comfortably fits the data shown. Columns that are much wider than the data should be avoided, as should columns that are narrow and appear to crowd the data. Although column headings and row labels will often be wider than the data shown, headings and labels can be written on more than one line to reduce the column width (for example, see the first and last columns in Table 8.5).
4. *Column alignment*: Within columns, digits of the same magnitude within numbers should be aligned vertically. Thus, although a word-processing program can centre all the figures in a column, this does not usually produce correct alignment. For example, the numbers 964.25 and 6.72 have different widths, so centring them will cause digits of the same magnitude not to appear in a straight vertical line. The rare exception is when the values happen to have the same number of digits.

   For non-numerical data, alignment should be consistent, but cell entries may be left-aligned or centred, depending on what seems appropriate. In Table 8.5, the names of the provinces were aligned on the left.
5. *Units*: At the top of a column, indicate the units of measurement for the data in that column, if applicable. The units should not be repeated in each cell. Dollars ($) and percentages (%) are units that are frequently repeated unnecessarily throughout tables. Note in Table 8.5 that the units for Land, Water, and Total areas were all in km^2, so one combined heading was made spanning the three columns. The strategic placement of lines makes it clear that there is a common characteristic in those three columns.

When you have finished creating a table, look over it from the perspective of a reader to see if the information is displayed effectively. Leaving a reader confused by what you have presented is never a good outcome, whether the information is written as text or displayed in a table.

Further Reading

Gustavii, B. (2008). *How to write and illustrate a scientific paper* (2nd ed.). New York: Cambridge University Press.

- Chapters 5–8 contain a good discussion of graph and table design, with examples.

Kosslyn, S.M. (2006). *Graph design for the eye and mind*. New York: Oxford University Press.

- This is a book devoted entirely to the presentation of information in graphs, with an emphasis on underlying principles of human visual perception and cognition. Examples include frequent side-by-side "Don't" and "Do" comparisons that explain how to improve visual characteristics of graphs so that they will be interpreted more easily and more accurately by readers.

Krohn, J. (1991). Why are graphs so central in science? *Biology and Philosophy, 6*(2), 181–203.

- This article examines critically the prominence, use, and significance of graphics in science.

Montgomery, S.L. (2003). *The Chicago guide to communicating science*. Chicago: University of Chicago Press.

- Chapter 9 is a brief though advanced guide to using figures and tables in scientific communication.

Nicol, A.A.M., and Pexman, P.M. (2003). *Displaying your findings: A practical guide for creating figures, posters, and presentations*. Washington: American Psychological Association.

- This is an extremely thorough examination of techniques for displaying information graphically. More than half the book—seven of twelve chapters—is devoted to describing the production of different types of graphs.

Pechenik, J.A. (2007). *A short guide to writing about biology* (6th ed.). New York: Pearson Longman.

- This guide contains very helpful information about preparing illustrative materials for effective presentation.

Wainer, H. (1984). How to display data badly. *The American Statistician, 38*(2), 137–47.

- This article is a fascinating analysis of 12 techniques for displaying data badly! Well worth reading.

9 Communicating with Maps

A good map is both a useful tool and a magic carpet to far away places.

—Author unknown

Key Topics

- What is the purpose of a map?
- What are the different types of maps?
- How is map scale related to detail?
- What are the characteristics of a good map?

Geographers would be lost without maps! Montello and Sutton (2006, p. 199) define cartographic maps as "graphic displays that depict earth-referenced features and data (geographic information), using a graphic space that represents a portion of the earth's surface." More generally, a map can be used to represent part of any surface (Boyd and Taffs, 2003, p. 47) or in fact to display a spatial or structural representation of any type of information (Montello and Sutton, 2006, p. 199). Maps are marvellous devices for presenting the results of research questions and for pointing out relationships that might otherwise be difficult to see. Indeed, maps can provoke more research questions than they resolve. This chapter discusses the importance of maps in geography and the environmental sciences and offers advice on presenting effective, informative maps.

What Is the Purpose of a Map?

At the broadest level there are two categories of cartographic maps (Slocum et al., 2009, p. 2): (1) general-reference maps and (2) thematic maps. As pointed out above, the underlying reason for creating a map is similar for both, but there is a difference in emphasis in the two categories of maps:

1. *General-reference maps* emphasize the location of multiple spatial phenomena. The purpose is to represent the features that are located on a part of the earth's surface. Inevitably this representation is simplified through the use of symbols, lines, and patterns. A **topographic map** is an example of a general-reference map, as is a highway map, although it pays particular attention to identifying roads and differentiating road types.
2. *Thematic maps* emphasize the spatial distribution of specific geographic attributes or data, also through the use of symbols, lines, and patterns. A selection of features on the earth's surface is normally included to show the spatial context and orient the reader. In contrast to general-reference maps, those contextual features are included as a design tool rather than being the purpose of creating the map. Examples of attributes shown on thematic maps are land cover, socio-economic variables, election voting patterns, disease occurrence, and transportation networks.

A thematic map has a more specific purpose than a general-reference map. Most maps that students will *create* within a geographically oriented degree program will be thematic maps. However, you can also expect to *use* general-reference maps, either in paper form (e.g., topographic maps) or on the web (e.g., Google Maps, Google Earth, or MapQuest). Slocum et al. (2009, p. 2) point out that the division between general-reference and thematic maps is actually one of convenience for categorizing different types of maps. A general-reference map in fact can be viewed as a thematic map that simultaneously shows multiple thematic elements at once.

What Are the Different Types of Maps?

There are many types or classes of maps; they differ in how they display spatial information (Robinson et al., 1994; Slocum et al., 2009). This section describes the types of maps that are most likely to be used by geography and environmental science students.

Dot maps

The **dot map** is a common way of showing both the spatial distribution and the quantity of a variable. Figure 9.1 is an example of a dot map. Dots are useful for showing the distribution of discontinuous or discrete data sets, such as population or oil wells. Each dot may represent a single occurrence of the unit being depicted in the area where the dot is placed, or some multiple of the unit.

On a simple dot map, each dot has the same value.

For example, a single dot might represent 1 grizzly bear den, 10 diseased trees, or 100 people. But the symbol does not have to be a dot; it could be any symbol. For example, a map of golf courses might use a golf-club symbol, while the distribution of airports might be shown with an airplane symbol. A **key** or **legend** must be included to indicate what the symbols represents.

On a simple dot map, all dots are identical, and they all represent the same quantity or value. A variation is to use different symbols to represent values of different magnitude (e.g., a dot for 10 people, a square for 100 people, a star for 1000 people).

Proportional dot maps

A proportional dot map is a variation on the simple dot map. Whereas a simple dot map uses dots (or other symbols) of identical size to represent

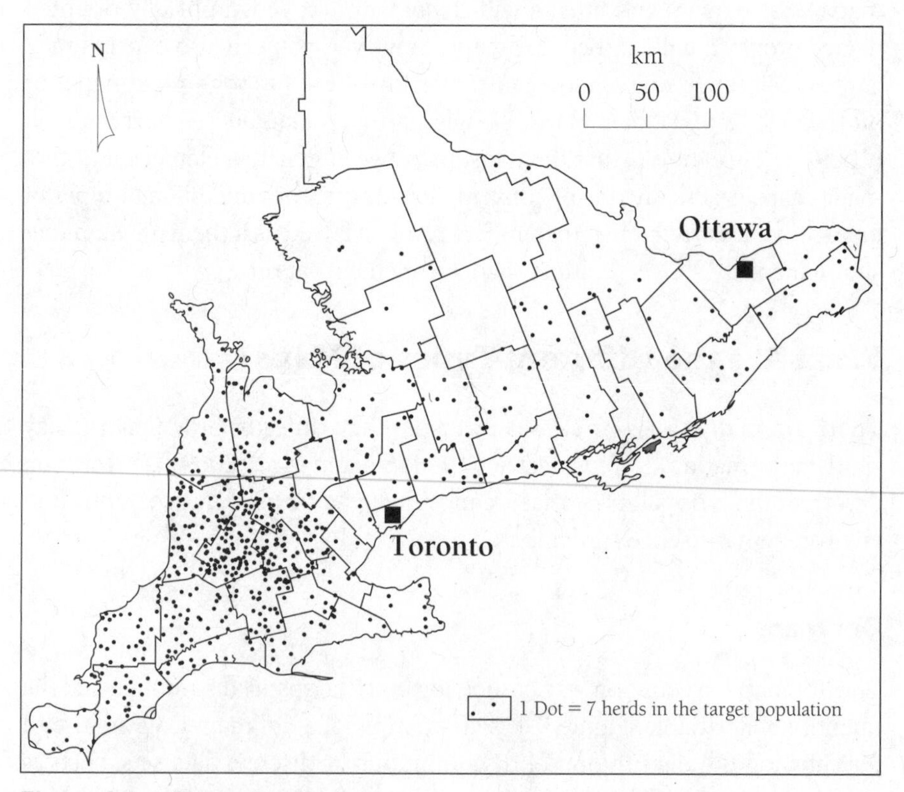

Figure 9.1 ◉ **Example of a dot map**

Swine farms in Southern Ontario, 2005

Source: Poljak et al. (2008).

numbers of some phenomenon, on a proportional dot map, as the name suggests, the size of each dot or symbol is directly related to the frequency or magnitude of the phenomenon represented. Figure 9.2 is an example of a proportional dot map constructed with circles.

If a circle is the symbol on a proportional dot map, the circles are created so that the *area* is proportional to the value of the geographical attribute at that location (this technique is known as *mathematical scaling*). This is achieved by making the diameter of the circle proportional to the square root of the number being illustrated. While it is also possible to create circles that have *diameters* proportional to the value, these dots are more likely to be misleading to the eye than circles based on area-value proportions. However, there is a complication, described by Slocum et al. (2009, p. 308): "Numerous studies have shown that the perceived size of proportional symbols does not correspond to their mathematical size; rather, people tend to underestimate the size of larger

On a proportional dot map, the size of each dot represents the magnitude of its value.

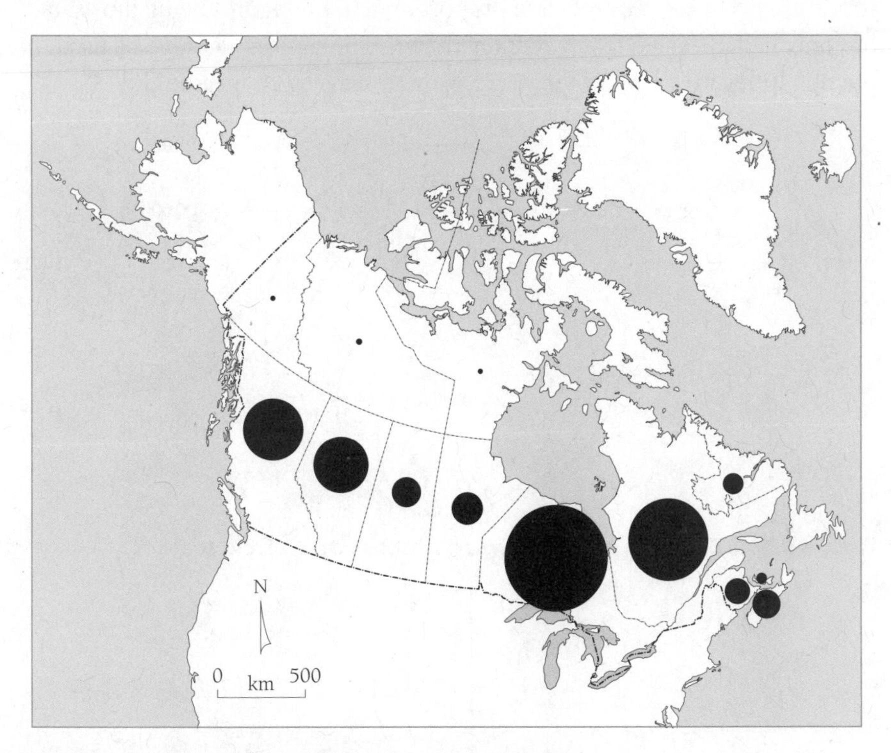

Figure 9.2 ◉ **Example of a proportional dot map**

Population of Canadian provinces and territories, 2008

Data source: Statistics Canada (2009).

symbols." This has led to the practice of using *perceptual scaling* to modify the area of symbols.

Figure 9.2 is a proportional dot map for the populations of the provinces and territories in Canada. It is important to include a **scale** for the phenomenon being mapped on a proportional dot map (see Figure 9.3). If circles are used, a series of circles of different sizes should be shown and labelled accordingly. The size scale is provided in addition to a linear scale for the map. Guidelines for presenting linear scale are given later in this chapter.

The technique of using proportional symbols may be usefully extended to represent two quantities (e.g., freight tonnages into and out of a region) in the form of a *split proportional circle*. To do this, follow the instructions above, but draw two semicircles at each point. Figure 9.4 is an example. The semicircle on one side might, for example, represent freight imports into a region (or births in a province), whereas the semicircle on the other side would represent freight exports (or deaths). The circles are made proportional to the figures being illustrated, but only half of each circle is actually drawn. The same scale must be used for the quantities in each half for visual comparisons to be meaningful.

Proportional pie charts are a useful way of showing magnitudes and proportions simultaneously.

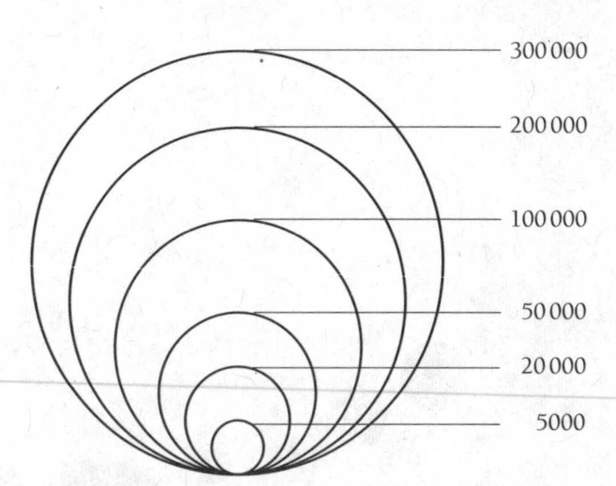

Figure 9.3 ◉ Example of a completed proportional circle scale

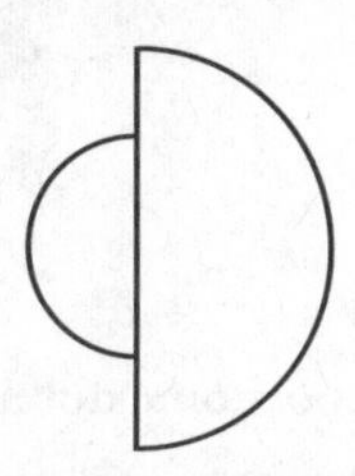

Figure 9.4 ◉ Example of a split proportional circle

Another way of displaying information on maps is as a *proportional pie diagram*, where the size of a circle will show the overall magnitude of a given value. The circle is then subdivided into different-sized segments, each of which represents some percentage of the total value. Each circle is a miniature pie graph (see the discussion of pie graphs in Chapter 8 for more information). For example, a map of housing types in the districts of Edmonton might show proportional circles indicating the total number of homes in each district, and these in turn might be internally subdivided to show the proportions of government housing, owner-occupied housing, rental dwellings, and so on in each of the mapped districts.

Choropleth maps

A **choropleth map** shows spatial distributions by means of cross-hatching, intensity of shading, or colours. Choropleth maps are commonly used to display classed quantitative data, e.g., rates, frequencies, and ratios such as marriage and divorce rates; population densities; birth and death rates; percentages of total population classified according to sex, age, ethnicity; and per-capita income. A choropleth map portrays the *aggregated* data for regions in discrete ranges or intervals, not continuously over space. *Aggregated* means that for each region on the map, a data interval is portrayed as being uniform throughout the region. Changes in value occur only at the borders of map regions if adjacent regions have different values. Choropleth maps may also be used to show categorical or qualitative data, such as types of soil or vegetation.

Figure 9.5 is a choropleth map that illustrates movements of people between census subdivisions in the Montreal metropolitan area between the 2001 and 2006 censuses. The map emphasizes a general pattern of migration away from the central area to the suburbs.

Choropleth maps are common for displaying aggregated data within regions, such as census tracts, electoral districts, provinces, territories, states, and countries. For example, results of federal elections are often summarized with a choropleth map in which each riding is coloured according to the winning party. Spatial patterns, such as clusters of ridings won by each party, may be revealed. To see the results of Canadian federal elections plotted on a choropleth map, visit the Elections Canada website (Elections Canada, 2013) at www.elections.ca. Note that since ridings with sparse populations spread over large areas dominate the map visually, inset maps are used to show areas of concentrated populations more clearly.

Choropeth maps are best suited for displaying four to six data classes.

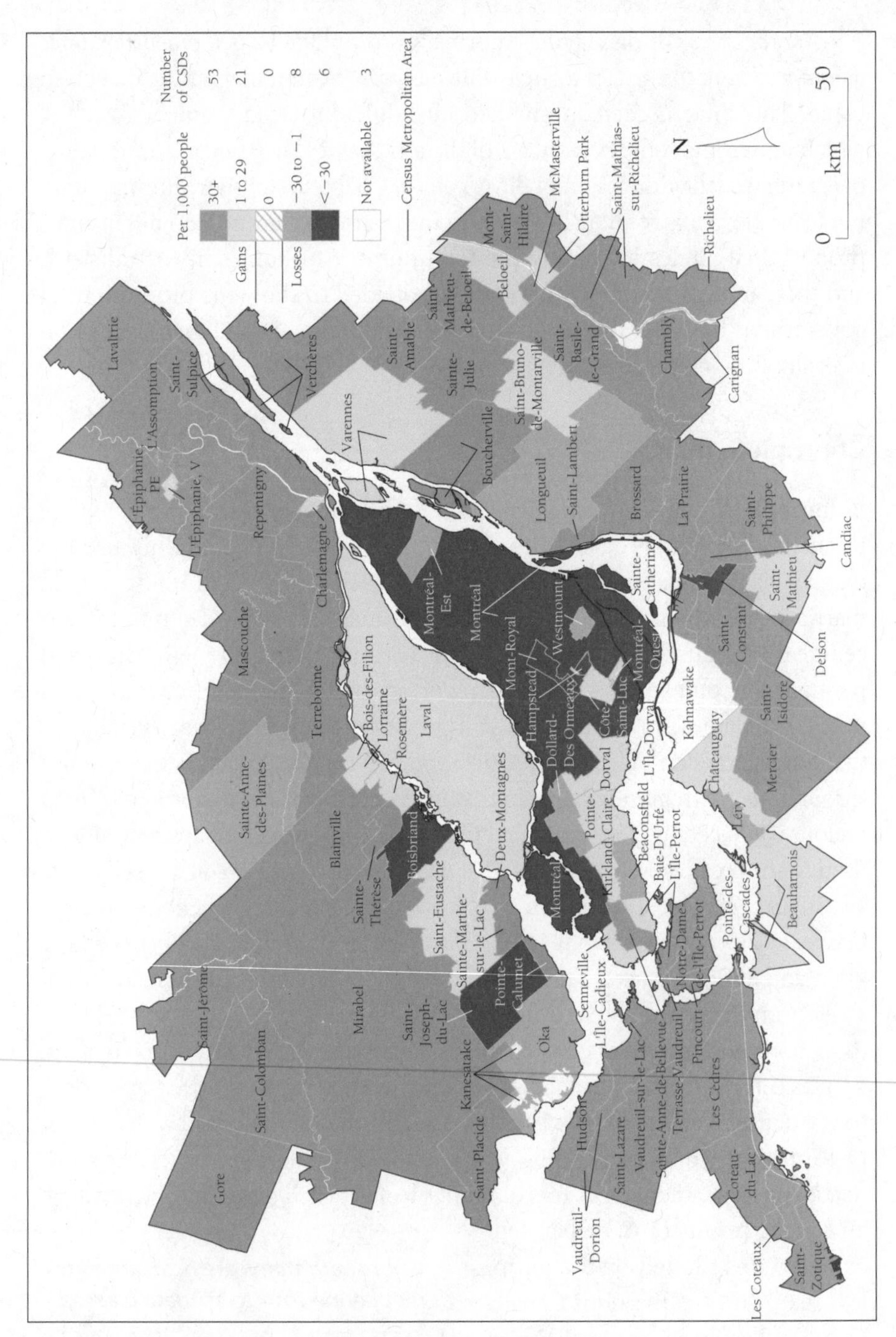

Figure 9.5 ◉ **Example of a choropleth map**

Migratory exchanges[1] within Montreal Census Metropolitan Area, 2001–2006

[1] Based on net migration rate

Source: Statistics Canada (2009).

If there are too many data classes on a choropleth map, the reader may find it difficult to visualize and interpret spatial patterns. Four to six classes is usually a good number. The process for selecting class intervals for a quantitative choropleth map is the same as that for choosing histogram class intervals described in Chapter 8. In both cases, a range of data must be divided into a specified number of classes. The most straightforward approach is to divide the range into equal intervals, but other methods may produce a better map display for a particular data set (see Slocum et al., 2009, Chap. 4).

Isoline maps

Isoline maps show sets of lines (**isolines**) that connect points that are known, or estimated, to have equal value. Many weather phenomena, such as temperature, precipitation, and pressure, are best displayed on isoline maps. Another frequently encountered form of map with isolines is the topographic map, depicting contours of equal elevation. Hydrographic charts show water depths with isolines supplemented by frequent point measurements. Table 9.1 lists common types of isolines and the variables they depict. Figure 9.6 contains isohyets (lines of equal precipitation) that reveal the pattern of annual precipitation in Mexico.

Isoline maps depict continuous data.

Other thematic maps

Numerous spatial phenomena can be portrayed on maps that, by definition, are thematic maps but are not specifically dot maps, choropleth maps, or isoline maps. Consider a map showing the migration paths of humans outwards from an east African origin over the past 20 000 years, with arrows indicating the directions of migration, and accompanying labels showing the approximate dates when the movements occurred. The network of arrows and labels could be drawn alone on a blank page; however, including the outline

Table 9.1 ● Some common isolines and the variables they depict

Isoline	Connecting Points of Equal
Isobar	Atmospheric pressure
Isotherm	Temperature
Isohyet	Precipitation
Contour Line	Elevation
Isobath	Water depth

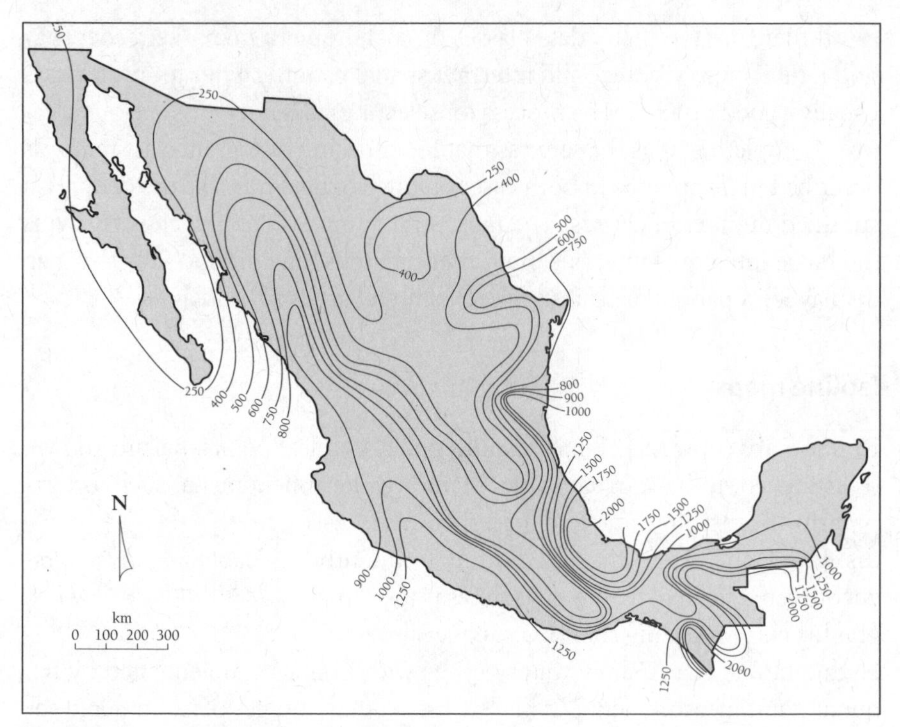

Figure 9.6 ◉ **Example of an isoline map**

Adapted from: University of Texas Libraries (2009).

of land areas at the continental scale would provide spatial context and make the map more effective for the intended purpose.

Maps published in newspapers and magazines (in print or in online versions) tend to be thematic in order to highlight a particular phenomenon and to complement the story. Similarly, many of the maps that a student might create to include in an essay or report are likely to have a particular theme. Box 9.1 describes two special types of thematic maps: topological maps and cartograms.

Box 9.1 ◉ Topological maps and cartograms

Near the beginning of any course on cartography, students are usually told that all maps contain distortions. This is because it is mathematically impossible to portray part of the earth's nearly spherical surface on a flat map without some type of distortion (Monmonier, 1996). A common goal in cartography, therefore, is to create a map that uses the kind of distortion that will be least disruptive to the purpose of the map. There are, however, maps that are produced to display geographic information more effectively by intentionally creating and taking advantage of distortions. These topological maps represent spatial adjacency or connectedness and are not as concerned with minimizing distortions.

Perhaps the most famous example of a topological map is the one showing the London Underground network (Transport for London, undated). In 1933, Harry Beck created a map that showed the positions of underground (subway) stations primarily in relation to other stations in the network. Routes were shown with straight line segments instead of the curved lines representing the true paths; distances between adjacent stations were standardized rather than scaled to reflect the actual distances, which varied. Beck realized that once passengers were at a departure station, the key piece of information they needed was the number of stops to the arrival station. The actual distance between stops was not very important. This allowed the network map to be portrayed with maximum efficiency. Relative location should not be abandoned completely on a topological map, however, because it should still represent a rough approximation of the true spatial distribution of features. Topological maps are common for displaying networks where the sequential order or relative position of items is more important to the user than their specific location.

Another interesting map is a cartogram, which scales the displayed size of individual administrative regions proportionally to the value of a variable of interest (Kimerling et al., 2005, p. 126). The regions are placed in their correct topological position (i.e., adjacency to neighbouring regions is preserved), but the size of each region on the map does not correspond to the region's land area. The unfamiliar patterns thus emphasize variations in the variable being shown while maintaining a sense of familiar geographical spatial relationships for the reader. Cartograms may be non-contiguous, pseudo-contiguous, or contiguous (Kimerling et al., 2005, p. 126).

An example of this type of distorted map is one that shows population by country. On a map that portrays countries with minimal distortions in comparative areas, Canada appears as the second-largest country in the world. But if the size of the Canada region on the map is scaled according to population, it becomes a "smaller" country, ranked thirty-seventh (Central Intelligence Agency, undated). Meanwhile a country with a smaller area but larger population has greater prominence: Bangladesh, for example, is ranked ninety-fourth in area but seventh in population.

Some fascinating new views of our world displayed with cartograms can be found on the web. Figure 9.7 is a contiguous cartogram showing world reserves of oil by country. Note that tiny Kuwait is the one-hundred-fifty-seventh-largest country by area (Central Intelligence Agency, undated) but the fourth-largest country on this map!

Topographic maps

Topographic maps are regarded as the most complete simplified representations of parts of the earth's surface. As will be discussed in the next section, the amount of detail shown is directly related to the scale of the map. A

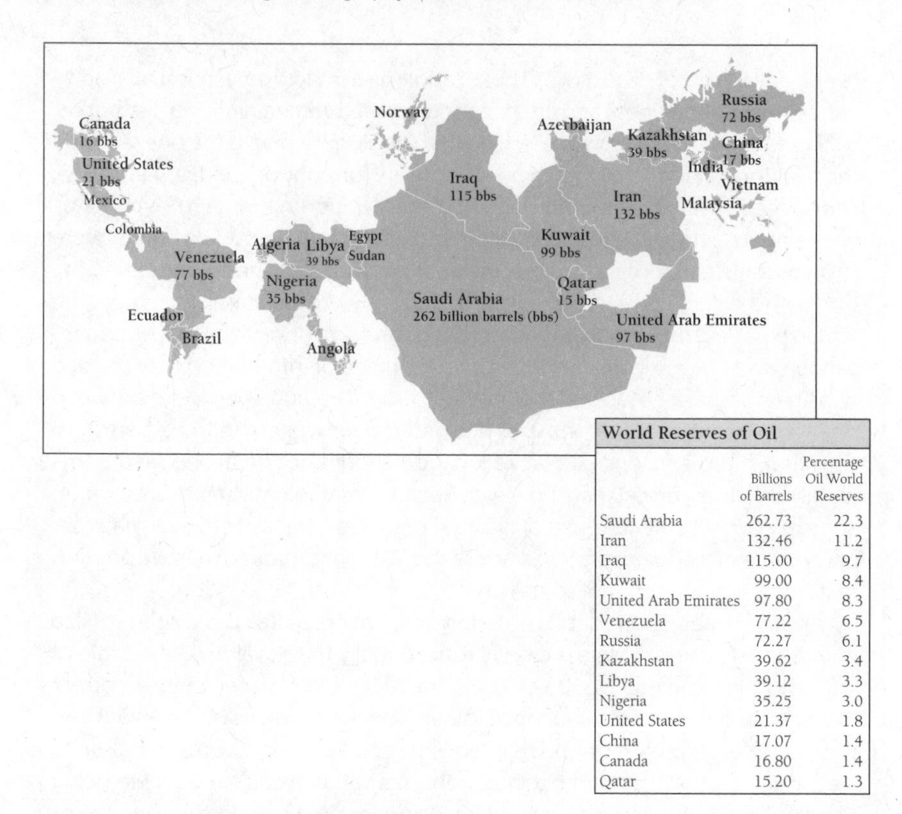

World Reserves of Oil

	Billions of Barrels	Percentage Oil World Reserves
Saudi Arabia	262.73	22.3
Iran	132.46	11.2
Iraq	115.00	9.7
Kuwait	99.00	8.4
United Arab Emirates	97.80	8.3
Venezuela	77.22	6.5
Russia	72.27	6.1
Kazakhstan	39.62	3.4
Libya	39.12	3.3
Nigeria	35.25	3.0
United States	21.37	1.8
China	17.07	1.4
Canada	16.80	1.4
Qatar	15.20	1.3

Figure 9.7 ◉ Example of a cartogram

World reserves of oil, 2004. Each country's size is proportional to the amount of oil it contains (oil reserves).

Adapted from: http://www.resilience.org/stories/2007-11-17/who-has-oil (accessed 23 July 2014). Data source: BP (2004) and U.S. Energy Information Administration (2013).

topographic map has a legend explaining the symbols, lines, and patterns used to represent the type and location of both natural and anthropogenic features, that is, "ground relief (landforms and terrain), drainage (lakes and rivers), forest cover, administrative areas, populated areas, transportation routes and facilities (including roads and railways), and other man-made features" (Centre for Topographic Information, 2007, unpaginated). A key feature of this type of map is the use of elevation isolines, or contours, which show the variations in elevation.

Topographic maps are general-reference maps with no single theme.

Many countries have topographic mapping agencies that produce a series of maps covering the country's land base. Each agency tends to produce maps with a consistent style, legend, and appearance. In Canada, the Centre for Topographic Information of Natural Resources Canada currently produces

National Topographic System (NTS) maps at two standard scales, 1:50 000 and 1:250 000.

Orthophoto maps

A type of map that is closely related to the topographic map is the **orthophoto map**. Here an ortho-rectified aerial photograph or satellite image forms the base of the map. On top of the photo base, a cartographer may add symbols, contour lines, or grid reference lines. Orthophoto maps are particularly useful for showing the actual landscape, whereas topographic maps are simplified representations of a landscape interpreted through a cartographer's eyes. Many of the maps of the northern regions of Canada in the National Topographic System are orthophoto maps, at least in earlier editions. Subsequent editions may be produced as topographic maps, but they require greater amounts of interpretation and cartographic work.

How Is Map Scale Related to Detail?

An inverse relationship generally exists between the **scale** of a map and the amount of detail it shows. To understand the reasoning behind this statement, we must understand the concept of *map* (or *cartographic*) *scale* and methods of displaying scale on a map.

Cartographic scale

Cartographic scale is the relationship between distance on the map and distance on the ground. That is, one unit of distance on the map represents *x of the same units* of distance in reality. Note the following:

- "*x*" is a numeric value that is specific to each map and is called the *scale factor*.
- Since, by definition, cartographic scale requires that the *same units* of linear measurement be used for distance on the map and on the ground, there is no need to specify particular units. For example, a measurement in millimetres (mm) or centimetres (cm) or metres (m) (or even inches) on a map can be used to determine the corresponding ground distance in millimetres or centimetres or metres (or inches).
- Because scale factors tend to be large numbers (e.g., between 1000 and 10 000 000), ground distances are usually converted from map units to more convenient units (e.g., on a 1:50 000 scale map, 1 mm represents

50 000 mm, or 50 m, in reality). While it would not be incorrect to tell someone that the actual distance between points was, say, 65 000 000 mm, it would be more useful to tell them that this distance was 65 km.
- People often quote the scale using the word *equals* instead of *represents* but this is not strictly correct (Monmonier, 1996, p. 7). One millimetre on the map does not "equal" 50 metres in reality; one millimetre represents a ground distance of 50 metres.

Cartographic scale on maps applies to distance, not area.

- The scale of a map applies to linear measurements only: that is, distance, not area. An areal scale could be used, but for the same map the areal scale factor would have to be a different number than the scale factor for the linear scale. Therefore, you should not attempt to substitute units of area directly for units of distance in the definition of cartographic scale above.

How to display the scale of a map

With certain exceptions (discussed later), the linear scale should be indicated on a map. There are two common techniques for informing the reader of a map's scale (*ratio scale* and *graphic scale*), and two other, less common techniques (representative fraction and verbal statement). Table 9.2 shows how the different forms of scale would be shown on a map.

- *Ratio scale*. The written definition of cartographic scale can be shown as 1: *scale factor* (e.g., 1:50 000) where the colon (:) means *represents*. Therefore, one unit of distance on the map *represents* 50 000 of the same units in reality.
- *Graphic scale* (also called *bar scale*). A line or rectangle (i.e., a bar, hence the alternative name) with known length is drawn on the map and labelled with the corresponding ground units that it represents.
- *Representative fraction*. This is identical to a ratio scale except that a division sign (/) is used instead of a colon (:), e.g., 1/scale factor (1/50 000). This emphasizes that a scale is a proportional relationship between the map and the area depicted.
- *Verbal statement*. The definition of cartographic scale, using the correct scale factor, is written directly on the map. It is called "verbal." Alternatively, the definition can be given with a selected map distance and the corresponding ground distance converted into convenient ground units. Table 9.2 shows examples for each alternative.

Table 9.2 ◉ Forms of cartographic scale portrayed on maps

Form of Scale	Portrayal on Map
Ratio	1:50 000
Graphic (bar)	0 200 km
Representative fraction	1/50 000
Verbal	One unit of distance on the map represents 50 000 of the same units in reality; or 1 cm on the map represents 50 m in reality

Ratio and graphic scales are used most frequently. Ratio scales have the advantages of showing the map scale with precision and, for more experienced users, allowing for quick recognition of the ground distances represented on the map (e.g., "It shows a scale of 1:100 000, so I know that 1 centimetre represents 100 metre."). The advantage of a graphic scale is that the user can visualize the ground distance represented by the length of the bar instantaneously, without doing any calculations. Because these scales each have distinct benefits for the user, it is common to see both forms on a single map.

Ratio and graphic scales are often both shown on a map.

A mental map is a unique case in which map scale is not necessarily constant across the whole map. Mental maps are described in Box 9.2.

Scale and map generalization

Map scales can be described on a continuum from large to small. Here it is instructive to go back to the idea of a representative fraction (e.g., 1/10 000, 1/100 000). The two scale factors in this example are 10 000 and 100 000. Notice that the fraction 1/10 000 is a *larger number* in decimal form (0.0001) than the fraction 1/100 000 (0.00001). Thus a 1:10 000 map is a *larger-scale* map than a 1:100 000 scale map. There is no specific dividing line between large- and small-scale maps. Compared to a 1:250 000 scale map, a 1:50 000 scale map is large-scale, but compared to a 1:5000 scale map, it is small-scale. World maps are definitely small-scale.

Consider two maps of identical physical size. The *larger-scale* map shows a *smaller part* of the earth's surface but can do so with *greater detail*. The *smaller-scale* map shows a *larger part* of the earth's surface but with *less detail*. The smaller-scale map is said to be generalized. For example, within a map

Box 9.2 Scale and mental maps

Usually the scale indicated on a map applies to all parts of the map; an exception is a mental map (or cognitive map), which usually has an inconsistent, or at least unreliable, scale. Another example is a map with an oblique perspective where scale varies. On such a map, it should be noted explicitly that the scale does vary across the map.

Mental maps are visualizations in the mind, but they can also be drawn in cartographic form, and they usually contain distortions in distances or size and shape of areas portrayed (Weston and Handy, 2004, p. 536; Kimerling et al., 2005, p. 4). Usually, the places that the map maker is more familiar with, or considers to be more important, are exaggerated in size while more distant or less familiar places are comparatively small. Similarly, the parts of the map that the maker is most familiar with tend to be more detailed. Mental maps are interesting from a psychological viewpoint because they provide some insight into the differences between perception and reality (Golledge and Garling, 2004; Weston and Handy, 2004).

Try sketching your mental map of a selected region to see how well your perception matches reality.

Transferring your mental map of an area to paper is an interesting exercise if you have not done it before. Try sketching a map of Canada with the outlines of the provinces and territories marked. When compared to a published map, does your mental map portray the provinces and territories with the correct relative sizes, or is there a size bias in favour of the region in which you live or with which you are more familiar? This exercise can be done for areas at various scales, such as countries, cities, and university campuses.

distance of 1 centimetre, a 1:250 000 scale map represents a ground distance of 250 metres. Within the same 1-centimetre map distance, a 1:50 000 scale map represents only 50 metres. The latter map can depict smaller features, or more details of the earth surface. On the smaller-scale map, specific features may be omitted or details may be generalized in order to produce a display that is clear and understandable. For the cartographer, map scale thus becomes a balance between two objectives: (1) displaying a region of interest within an available space and (2) showing an appropriate level of detail.

What Are the Characteristics of a Good Map?

Cartography is considered by many to be an art, and what is pleasing in art is something that to many people cannot be distilled into a list of standard principles. Nevertheless, there are some basic principles that guide map design and that should be considered by anyone creating and presenting a map, whether using computerized or manual methods (Box 9.3). With writing, rules can be broken, but it is usually better to have a firm grasp of the rules before one attempts to break them. The same applies to maps: you should understand the reasoning behind good practices before producing maps that do not accord with those practices. Statements that begin with "All maps should . . . " can therefore be seen as guiding principles to which exceptions are possible.

Box 9.3 ◉ Computerized versus manual map creation

Before the age of computers, maps were drafted by hand (see the web for examples of classical maps drawn hundreds or thousands of years ago, e.g., National Geographic Society, undated). Today professional cartographers use computers with specialized mapping programs, such as AutoCAD; geographic information systems, such as ArcGIS; or general graphics programs like Adobe Illustrator or CorelDRAW. Just as word processing has replaced handwriting for the presentation of written work at the undergraduate level, there is a growing expectation that maps will be created digitally.

Using a computer can help to create maps that are more consistent in appearance than those drawn by hand. Cartographic or GIS software programs automate many of the steps outlined in this chapter, making map production faster and easier (Oliver, 2000). Be careful, however, not to rely on the software alone to create a map. The user's input is still critical for designing an effective map. Many default characteristics can be customized to improve a map's appearance.

Computerized maps designed to be viewed on a computer or electronic devices offer the cartographer more options for interactivity than a basic map printed on paper (see, e.g., Cartwright, et al., 2007 and Peterson, 2008). Such features, which are beyond the scope of this book, require specific instruction such as that provided in a cartography course. In this book, the focus is on guidelines for the design of basic maps that a student could consider incorporating into an essay or report.

Designing a map is different from actually drawing it. The specific techniques of "drawing" may be different, but the principles of good map design that applied to hand-drawn maps still apply to maps drawn with computer software. Good map design requires thought, planning, and care so that the information shown on the map is presented clearly. Clarity, legibility, and inclusion of critical map elements are important on all maps, regardless of the method of creation. MacEachren (1995) emphasizes the role of human perceptual processes in seeing and understanding maps, and it is useful to consider how these processes should influence map design.

High-quality maps are displayed in both *Canadian Geographic* and *National Geographic*. These publications employ professional cartographers who produce some exquisite maps that are works of art as well as effective forms of communication. A student who examines and tries to emulate the design and presentation of those maps will likely produce top-notch work. At the same time, some published maps are of poor quality. When you look at a map, ask yourself whether its design and display are effective. If they are not, you may still be able to learn from it and avoid repeating its creator's mistakes.

Study the characteristics of published maps to improve the appearance of maps you create.

Standard map elements

Maps should have all of the following elements, with the exceptions that are noted:

- *Title*. As with a graph or diagram, a title is important to indicate the map's purpose or features.
- *North arrow*. An arrow pointing north orients the map reader. By convention, north is often at the "top" of a map, but with the inclusion of a north arrow this is not a requirement. North arrows are often omitted when it is expected that an average user will be sufficiently oriented by his or her knowledge of the region depicted (as in the case with continental or world maps). On these smaller-scale maps, the direction toward north is often not consistent on all parts of the map anyway, because of distortion, so the omission of a north arrow is reasonable. In place of a north arrow, cartographers may show latitude and longitude with marks on the map borders or with grid lines. The marks or lines must be labelled (and the hemisphere named, usually by showing N, S, E, or W beside the latitude and longitude values).
- *Scale*. The user needs to know the relationship between map distances and ground distances, as described above.
- *Border*. A map should have a border to show the limits of the region being depicted unless the region has a complete boundary. For example, since the island of Newfoundland has a complete and identifiable boundary, a map border might be omitted. However, a map of a small part of central Newfoundland would require a border to show the limits of the mapped region. Even if the region does have a complete boundary, it is often wise to include a border to separate the map from other figures or text.

Other elements, such as information about *projection*, *datum*, and *ellipsoid*, may be included on maps but are not as standard as those in the preceding list. Maps created in GIS software may include these elements because the information is readily available and the map user will then be fully informed about the details of its creation. An element often included on maps of larger regions, such as continents or the world, is the name of the projection. This is done on such maps because the choice of projection is much more evident in the size, shape, and relative position of features than on a map of a smaller region.

Map design elements

Good map design is often determined as much by individual preferences as by standard guidelines. That being said, there are resources for learning about map design (e.g., Krygier and Wood, 2005; Peterson, 2009). Here are some important things to consider when designing a map:

- *Clarity*. Avoid a cluttered appearance that detracts from the map's main purpose. Reduce clutter by including only enough features and labels to orient the user and convey the required information.
- *Contrast of patterns and colours*. Make sure patterns and colours are sufficiently distinct. A common problem with choropleth maps is that there is not enough contrast between different shades or colours. This usually occurs because too many classes are portrayed or because the contrast between minimum and maximum values is too small. Patterns can also be difficult to distinguish if too many are used.
- *Lines and symbols*. Choose line styles and symbols that are visually distinct. Maps are often confusing because the same line style is used for roads and coastlines. Use symbols that are neither too small to be distinguished clearly nor so large that they cover too much space.
- *Labels and lettering*. Use lettering that can be read easily at a glance. Labels should be close to the corresponding feature and of a size that matches the importance of the feature.
- *Inset map*. It is often useful to include an inset map in the corner of the main map to orient the user better than the main map does alone. For example, if you have created a map of central Newfoundland, include a small inset map that shows the whole island of Newfoundland with the mapped central area marked. This will help to orient users not familiar with central Newfoundland.

Consider including an inset map to help orient the user.

Remember that a map has a purpose. A better-presented map is more useful for helping a person understand its intended purpose. While it may take extra effort to improve the appearance of a map produced using the default settings in a software package, a customized map may be significantly more appealing visually, more informative, and therefore more effective.

Further Reading

Kimerling, A.J., Muehrcke, P.C., and Muehrcke, J.O. (2005). *Map use: Reading, analysis, and interpretation*. Madison, WI: JP Publications.

- As the title suggests, this book is about *using* maps, not constructing them; however, it is a highly recommended reference for learning about maps and map design.

MacEachren, A.M. (1995). *How maps work*. New York: Guilford Press.

- This is a good source for someone looking for a deeper background understanding of the perceptual processes that determine how a user sees and understands a map. With this knowledge, better map design can be achieved.

Monmonier, M. (1996). *How to lie with maps* (2nd ed.). Chicago: University of Chicago Press.

- This book is a very interesting presentation of the various ways in which maps do not represent reality with complete accuracy. A lie may be either an inevitable consequence of the mapping process or the result of a more deliberate attempt to deceive. By explaining how to "lie" with maps, this book makes good mapping practice clearer.

Peterson, G.N. (2009). *GIS cartography: A guide to effective map design*. Boca Raton, FL: CRC Press.

- With the move toward the use of GIS packages for creating maps, this compact volume focuses on issues of designing effective maps in a GIS environment.

Slocum, T.A., McMaster, R.B., Kessler, F.C., and Howard, H.H. (2009). *Thematic cartography and geovisualization* (3rd ed.). Upper Saddle River, NJ: Pearson Education.

- This book provides an extensive explanation of thematic mapping, including background cartographic issues, techniques for constructing different types of thematic maps, and geovisualization.

10 Preparing and Delivering an Oral Presentation

All the great speakers were bad speakers at first.

—Ralph Waldo Emerson

An orator is someone who says what he thinks and feels what he says.

—William Jennings Bryan

Key Topics

- Why is public speaking important?
- How to prepare for an oral presentation
- How to deliver a successful oral presentation
- How to cope with questions from the audience

Why Is Public Speaking Important?

Public speaking serves valuable intellectual, vocational, and professional functions.

Although talking comes naturally to most of us, many people continue to rate public speaking as one of their top fears. And believe it or not, professors do appreciate the fact that public speaking is intimidating to many students. While some professors have grown accustomed to speaking before a large audience, many still feel some trepidation about speaking in front of an unfamiliar class, at a professional gathering, at a community meeting, or even at a wedding party. They do understand the sleepless nights, sweaty palms, pounding heart, dry mouth, and weak legs that sometimes precede an oral presentation. So, when you are asked by your professor to give a prepared **oral presentation** in class, it is unlikely that the assignment has been chosen lightly. Professors usually have three fundamental objectives in mind when they ask you to give a presentation in your geography or environmental science class.

First, preparing for and delivering an oral presentation encourages you *to organize your ideas, to construct logical arguments, and to otherwise fulfil the objectives of a post-secondary education* (for a discussion, see Jenkins and Pepper, 1988).

Second, your professors also have your *career interests* in mind. Many of the jobs in which geographers and environmental scientists find themselves require them to make public presentations. While business and educational leaders acknowledge oral communication and public-speaking skills to be among the most important abilities graduates can have (see, for example, Hay, 1994), a number of international surveys have found that these skills are also among the most poorly developed. Consequently, your future employers are likely to be impressed if you can point out to them that, through your degree, you gained the experience of giving several multimedia presentations to live audiences, particularly if you can explain that you have used each of those opportunities to refine your presentation skills.

Third, the ability to speak effectively at conferences and other professional or community gatherings is *a critical component of effective geographical or environmental science practice* (see also the discussion on public engagement in Chapter 11). If we are to make contributions to our significant work in areas such as environmental justice, global warming, and social rights, we must be able to share the results of our work in such venues (Hay et al., 2005, p. 159). Given the importance of these skills, you will probably not be spared from having to give one or several "public presentations" of about 10 to 20 minutes duration throughout your studies—despite any fears of speaking that you might have. The "mechanics" of giving such an oral presentation are outlined in this chapter, which is divided into three main parts. The first deals with the essentials of preparing for the presentation; the second, with delivering the presentation; and the third, with coping with questions from the audience.

Throughout the chapter, reference is made to the use of Microsoft **PowerPoint** software. This widely used aid may help you to produce a more effective presentation for your audience and, if applicable, high-quality handouts. PowerPoint is mentioned specifically because of its ubiquitous presence on institutional computer networks, but it is not the only software program available for creating presentations. A cautionary note—applicable to PowerPoint, but even more so if you use a different program—is that you should find out in advance what programs will be available in the venue where you will be making your presentation. You do not want to show up with a presentation in a format that is not supported by the available software.

The discussion that follows is not intended to be a prescription for a perfect oral presentation: Instead, it offers guidelines to help you prepare for and deliver your first few "public speeches." With experience you will develop your own style—that is, an effective form of presentation that you are comfortable with even if it transgresses some of the guidelines discussed here. Practice will help you to develop your own approach, but you may also want to be attuned to your professors and to other people who give oral presentations that you attend. Pay attention to the form and manner of their delivery. Try to identify those devices, techniques, and mannerisms that you believe add to a presentation (or detract from it—not all speakers, even professors, are perfect). Apply what you learn to your own presentations. You may also find it helpful to look over Box 10.1 to get some idea of what contributes to a successful oral presentation.

Try to develop your own style of public speaking, but be aware of your audience's expectations.

Box 10.1 ◉ Assessment form for an oral presentation

Student Name: Grade: Assessed by:

The following is an itemized rating scale of various aspects of a formal oral presentation. Sections left blank are not relevant to the presentation assessed. Some aspects are more important than others, so there is no mathematical formula connecting the scatter of ticks with the final percentage for the presentation. Ticks in either of the two boxes left of centre mean that the statement is true to a greater (outer left) or lesser (inner left) extent. The same principle applies to the right-hand boxes. If you have any questions about the individual scales, comments, final grade, or other aspects of this assignment, please see the assessor indicated above.

First impressions

Speaker appeared confident and purposeful before starting					Speaker appeared to lack confidence and purpose
Speaker's personal appearance of high standard					Speaker did not pay attention to personal appearance
Speaker attracted audience's attention from the outset					Speaker did not attract audience's attention

Presentation structure

Introduction

Title and topic made clear					Title or topic not made clear

Continued

Purpose of the presentation made clear					Purpose not made clear
Organizational framework made known to audience					Audience not informed about organizational framework
Unusual terms defined adequately					Unusual terms defined poorly or not at all
Body of presentation					
Main points stated clearly					Main points not stated clearly
Sufficient information and detail provided					Insufficient information or detail provided
Appropriate and adequate use of examples or anecdotes					Inappropriate or inadequate use of examples or anecdotes
Discussion flowed logically					Discussion was confusing or illogical
Conclusion					
Ending of presentation was signalled adequately					Ending of presentation was abrupt
Main points were summarized adequately					Main points were not summarized
Ideas were brought together well					Ideas were not brought together
Final message clear and easy to remember					No clear final message
Coping with questions					
Whole audience was searched for questions					Parts of audience were ignored during question period
Questions were handled adeptly					Difficulty was encountered in handling questions
Full audience was addressed with answers					Only the questioners were addressed during answers
Speaker maintained control of discussion					Speaker lost control of discussion

Delivery

Speech was clear and audible to entire audience					Speech was unclear or inaudible to parts or all of audience
Speaker was engaged and showed enthusiasm					Speaker was not engaged or lacked enthusiasm
Presentation was directed to all parts of audience					Presentation was directed only to parts of audience
Eye contact with audience was maintained throughout presentation					Eye contact with audience was missing or inconsistent
Presentation length was close to time limit					Presentation was excessively below or above time limit
Good use was made of time without rushing at end					Poor use of time led to rushing at end
Pace was comfortable for audience					Pace was too fast or too slow

Visual aids and handouts (if relevant)

Visual aids were well prepared					Visual aids were poorly prepared
Visual aids were visible to entire audience					Visual aids were not visible to entire audience
Speaker was familiar with presentation equipment					Speaker showed lack of familiarity with presentation equipment
Effective use was made of visual aids					Visual aids were used ineffectively
Handouts were well prepared and useful					Handouts were poorly prepared or lacked relevance

Assessor's comments

____

____

____

____

How to Prepare for an Oral Presentation

You cannot expect to speak competently off-the-cuff on any but the most familiar topics. Effective preparation, which is crucial to any successful presentation, should begin some days (at least) in advance of the actual event and certainly not just the night before. Give yourself plenty of time to revise and rehearse. But before you can prepare your presentation, you must do several things first.

Establish the context and goals

- *Who is your audience?* Target the presentation to the audience's characteristics, needs, and abilities (Gurak, 2000). The ways in which a topic might be developed will be critically influenced by the background and expertise of your listeners. Consider, too, how big the audience will be since this may affect the style of presentation. For example, it will be more difficult to use an interactive style of presentation effectively with a large audience than with a small one.
- *Why are you speaking?* The style of presentation may also differ depending on your purpose. The purpose may be to present information, to stimulate discussion, to present a solution to a problem, or perhaps to persuade a group of the value of a particular view or course of action. Depending on the presentation's purpose, you may have to alter its style and content.

> ***To give a successful oral presentation, make sure you know why you are speaking, where, to whom, and for how long, but, above all, work out what your central message will be.***

- *What is your subject?* Be sure that your subject matches what the audience is expecting to hear. A mismatch may upset, bore, or alienate your audience. A clear sense of purpose will also allow you to focus your presentation more clearly.
- *Have you done your research?* Keeping in mind the purpose of your presentation, gather and interpret relevant and accurate information. Make a point of collecting anecdotes, up-to-date statistics, or cartoons that might make your presentation more appealing, convincing, and colourful.
- *How long will you speak?* Confirm how much of your allotted time is allocated for the oral presentation and how much is intended for questions from the audience. Avoid the embarrassment of being asked to conclude the presentation before you have finished, or of ending well short of the scheduled length. Fitting your presentation to the time allotted requires both planning and practising in advance.

- *Where are you speaking?* Student presentations for courses are usually given in the regular classroom, so you will probably be familiar with the place where you will be presenting. If not, or if your presentation is not part of a course, it is wise to visit, in advance, the venue where the presentation is to be held. The room and its layout can have an effect on the presentation's formality and speed, the audience's attentiveness, and the types of audiovisual aids that can be employed. Check, for example, to see if the presentation is to be given in a large room, from a lectern, with a microphone, to an audience seated in rows, and so on.
- *Is your software program installed?* If you plan to use a presentation software program, make sure it will be available, particularly if it is not a standard program such as PowerPoint. Sometimes presenters assume they will be able to run the presentation off their own laptop instead of the computer in the venue, such as when they have material in a different software program. If you think this is absolutely necessary, you should confirm in advance that doing so will be permitted (conference organizers often disallow this practice for efficiency and to avoid equipment glitches). Also make sure that the necessary connection hardware will be available, and that it will not disrupt the flow of the event in order to get set up. Remember, the audience wants to hear what you have to say, not to watch you fumbling with cables, plugs, and adjustments to display settings. Using the standard equipment and software provided at the venue is always the simplest and best option whenever possible.

Organize the material for presentation

- *Eliminate superfluous material.* If your presentation is to be based on a paper you have already written, be aware that you will not be able to say everything you have written. Carefully select the main points and decide on the strategies to communicate them as clearly and effectively as possible. Courtenay (1992, p. 220) makes the following suggestions:
    - List all the things you know or have found out about your subject.
    - Eliminate all those items you think the audience might already know about.
    - Eliminate anything that is not important for your audience to know. Keep doing this until you are left with the most important and dynamic points. These should not already be known to your audience, and they should be interesting and useful to them. These points should form the basis of your presentation. Stettner (1992, p. 226) makes a very good argument for organizing an oral

presentation around no more than, and no fewer than, three main points.

- *Give your presentation a clear and relevant title.* Just as with an essay, an audience will be attracted to, and informed by, a good **title**. Be sure that your title clearly identifies the presentation's subject matter.
- *Choose the right framework.* Ensure that the *organizational framework* helps make the point of your presentation clear (see Box 10.2). For example, if your main aim is to discuss potential solutions to homelessness in Saskatoon, it would probably be useful to mention social security policies in Saskatchewan as a prelude to that, but not very useful to spend most of the presentation discussing the historical development of those policies.

Box 10.2 Some organizational frameworks for an oral presentation

Most presentations seem to have one of the following five organizational frameworks:

1. *Chronological*: e.g., changes in acid precipitation levels in eastern North America in the past 50 years;
2. *Scale*: e.g., overview of national responses to climate change followed by detailed examination of responses in a particular area;
3. *Spatial*: e.g., a description of Japan's trading relations with other countries of the Pacific;
4. *Causal*: e.g., implications of financial deregulation for the Canadian insurance market;
5. *Order of importance*: e.g., ranked list of solutions to the problem of homelessness in Saskatoon

Structure your presentation

A great oral presentation will have a structure that has the following elements:

- Opens with a clear, memorable statement;
- Concentrates on a small number of key points;
- Is concise;
- Is "signposted";
- Ends with a clear, memorable statement that is consistent with the opening;

(Knight and Parsons, 2003, p. 164).

Let us look at the structure of an oral presentation in more detail. In most cases it will have an introduction, a body, and a conclusion—just like an essay. The introductory and concluding sections are very important. Indeed, about 25 per cent of your presentation ought to be devoted to the "beginning" and "end" (combined). The remaining time should be spent on the body.

Introduction

- *Make your rationale for speaking and your* **conceptual framework** *clear.* This gives the audience a basis for understanding the ideas that follow. In short, let the listeners know what you are going to tell them. Box 10.3 explains how to do this.
- *Capture the audience's attention from the outset.* Do this with a **rhetorical question**, relevant and interesting quotations, amazing facts, an anecdote, or a startling statement. Avoid jokes unless you have a real gift for humour.
- *Make the introduction clear and lively.* First impressions are very important.

Try to capture your audience's attention from the very start of your presentation.

Box 10.3 ◉ Introducing an oral presentation

- *State the topic.* "I am going to speak about . . ." Do this in a way that will attract the audience's attention.
- *State the aims or purpose.* Why is this presentation being given? Why have you chosen this topic? Why should the audience listen?
- *Outline the scope of the presentation.* Let the audience know something about the spatial, temporal, and intellectual boundaries of the presentation. For example, are you discussing Aboriginal perspectives on the environment? Or offering a geographer's view of British financial services in the past decade? Some people love to point out gaps in your work's coverage, so head them off by making clear what you are and are not going to cover.
- *Provide a plan of the discussion.* Let the audience know the steps by which you will lead them in your presentation and the relationship of each step to the others. It is useful to show the audience an outline of your intended progression.
- *Give the audience enough time to absorb information.* A common error that presenters make is to show the audience information on the screen without leaving enough time for them to make sense of it (Alley, 2003, p. 66). Avoiding this error is particularly important during the introduction, when you are still trying to catch the audience's attention.

Body

- *Construct a convincing argument supported with examples.* Remember, you are trying to present as compelling a case as possible in support of your findings.
- *Consider the flow of the presentation.* Ensure there is a "fluid logic between your main points" (Montgomery, 2003, p. 172).
- *Limit discussion to a few main points.* Do not make the mistake of trying to cover too much material for the available time. Making a few points with depth and supporting evidence is better than making a larger number of points superficially.

Limit your presentation to a discussion of key points, and make clear the relationship of each to the path you are following.

- *Present your argument logically, precisely, and in an orderly fashion.* Try producing a small diagram or short list that summarizes the main points you wish to discuss. Use this as a basis for constructing your presentation. It might also be turned into a useful handout or PowerPoint slide for your audience.
- *Accompany points of argument with carefully chosen, colourful, and correct examples and analogies.* It is helpful to use examples built upon your audience's experiences. Analogies and examples clarify unfamiliar ideas and bring your argument to life.
- *Connect the points of your discussion with the overall direction of the presentation.* Remind the audience of the path you are following by relating the points you make to the framework you outlined in the introduction. For example, "Moving on to the third of my main points, I will discuss . . . "
- *Restate important points.* Reminding the audience of key points will reinforce them in their minds.
- *Personalize the presentation.* This can add authenticity, impact, and humour. For example, in discussing problems in administering a household questionnaire survey, you might recount an experience of being chased down dark suburban streets by a large, ferocious dog. Avoid overstepping the line between personalizing and being self-centred by ensuring that the tales you tell will actually help the audience become engaged and understand your message.

Conclusion

- *Cue the conclusion.* Phrases like "To conclude . . ." or "In summary . . ." have a remarkable capacity to stimulate an audience's attention.
- *Bring ideas to fruition.* Let the audience know exactly what the take-home message is. Restate the main points in words other than those you used

earlier in the discussion, develop some conclusions, and review the implications. Connect your presentation with its wider context.

> ***When you tell your audience you are about to finish, you get their attention!***

- *Tie the conclusion neatly together with the introduction.* The introduction noted where the presentation would be going. The conclusion reminds the audience of the content and dramatically marks the arrival at the foreshadowed destination.
- *Make the conclusion emphatic.* Do not end with a whimper! A good conclusion is very important to an effective presentation because it reinforces the main ideas or motivates the audience (Eisenberg, 1992, p. 340). For instance, if you have been stressing the need for community involvement in reducing greenhouse gas emissions, try to "fire up" the members of the audience so that they feel motivated to take some action of their own.
- *End the presentation clearly.* Saying "thank you," for example, makes it clear to the audience that your oral presentation is over. Avoid letting your speech drift away to an inaudible level. Often this creates an awkward pause followed by a weak confirmation from the speaker that he or she is in fact finished. Until you give a clear signal that you are finished, the audience will tend to wait quietly in the belief that there might be more coming. Given the right signal, the audience will usually reward you with a round of applause that you (hopefully) deserve, given the time and effort you put into preparing the presentation.

Prepare your text and aids to delivery

Different people have different preferences when it comes to writing a presentation "script." Some opt for a full script, others for brief notes, and others for graphic images such as a flow diagram. The same model is not suitable for everyone, so heed Montgomery's (2003, p. 171) advice to "design and write out your presentation in a manner you feel comfortable with." That said, there are a few points worth considering.

Notes

- *Prepare well in advance.* Mark Twain is reported to have said, "It usually takes more than three weeks to prepare a good impromptu speech" (in Windschuttle and Elliott, 1999). Twain may have overstated the case a little, but it is fair to consider the presentation the tip of the iceberg and the advance preparation the much larger submerged section.

- *Prepare a presentation, not a speech.* In general, you should avoid preparing a full text to be read aloud. A presentation that is simply read aloud is often boring and lifeless because it tends to sound stilted and unnatural, and there is insufficient contact with the audience. (It is conceded that there is not a universal prohibition on reading a full text aloud since this is done at some conferences, most commonly in the humanities.) If you must prepare a text to be read, remember to keep it simple and logical. Because your audience will hear your presentation only once, it must also "be very well organized, developed logically, stripped of details that divert the listener's attention from the essential points of the presentation . . ." (Pechenik, 2007, p. 259).
- *Write for presenting, not reading.* Although most people speak at between 125 and 175 words per minute (Dixon, 2004, p. 100), a good speed for formal presentation delivery is only about 100 words per minute (Montgomery, 2003, p. 171). So, if you feel absolutely compelled to write a script, you will know that a 10-minute presentation will require you to prepare about 1000 words. Keep sentences short and simple. A text that would be suitable in a research paper is not appropriate as an oral presentation without modifying the language (Beins and Beins, 2008, p. 210). If the presentation is based on a previously written paper, the text should at least be "translated" from written to oral style. The language should be informal, but you should not use slang or other conventions of speech that you might use at a bar or write in a text message to a close friend.

If you use a script for an oral presentation, make sure it is designed for speaking, not reading.

- *Prepare personal memory prompts.* These might take the form of clearly legible notes, key words, phrases, or diagrams to serve as the outline of your presentation. Put the prompts on cards, and number them—just in case you drop them! With PowerPoint, consider using the "View Notes Page" option. This useful option allows you to prepare and print out a set of speaker's notes for each slide in your show.
- *Revise your script.* Put your presentation away overnight or for a few days after you think you have finished writing it. Come back to the script later, asking yourself how to sharpen the presentation.

Handouts

- *Consider preparing a written summary for the audience.* In general, an oral presentation should be used to present the essence of some body of material. You might imagine the presentation to be like a trailer for a forthcoming movie that presents highlights and captures the imagination.

If members of the audience want to know more, they should come along to the full screening of the film, i.e., read the full paper. Depending on the circumstances, it may be helpful, therefore, to prepare a full copy or summary of the paper on which the presentation is based to be distributed to the audience. You may also consider preparing a condensed version or a handout with a few key diagrams, the main points of your presentation, and your contact information.

PowerPoint includes among its print options a function that allows you to prepare handout copies of images you plan to use during your presentation. With the aid of such a document, the audience is better able to keep track of the presentation and you are freer to highlight the central ideas and findings instead of spending valuable time on detailed explanations.

Visual aids

At the professional level, digital projectors have supplanted overhead projectors and 35 mm slide shows for displaying visual material to accompany a presentation. Furthermore, PowerPoint is the "nearly universal standard" presentation software in use (Anholt, 2006, p. 73). Its widespread availability, along with installation of the required projection equipment in classrooms and lecture halls, means that PowerPoint is commonly used at the undergraduate and graduate levels, too. Box 10.4 highlights some key do's and don'ts when using presentation software, such as PowerPoint, and further advice on preparing an oral presentation is given following it.

Box 10.4 ● Effective use of presentation software

This box considers characteristics of visual aids in general that will help to improve the audience's understanding and appreciation of your work. Here is a summary of key pointers to make the most effective use of presentation software such as PowerPoint.

Do:

- Create a presentation that unfolds with a logical flow of information. Have a clear introduction, body, and conclusion.
- Make sure text on the screen is legible from anywhere in the venue, not just at the front. It is likely you will be closer to the screen than anybody in the audience, and material clearly legible to you may not be so clear to someone seated at the back of the room.
- Consider pausing before moving on to the next slide to give the audience time to absorb and process the information before being confronted with

Continued

new information.

- Use the software to create a more interesting visual presentation to accompany your oral delivery. For example, various illustrative devices can help the audience understand your material much more easily than either reading text or listening to a complicated explanation.
- Project your voice to the audience—this is difficult to do if you are facing away from audience members and speaking toward the screen.

Don't:

- Read verbatim off the screen. Text on the screen should summarize and emphasize your main point but be supplemented with information given orally—it is, after all, called an oral presentation!
- Speak about material that is unrelated to what is on the screen—it is difficult to listen and read effectively at the same time, and the audience will be confused about whether they should listen to you or make sense of what is on the screen.
- Include more visual or graphic detail on a slide than you have time to describe or explain clearly.
- Use graphic animation features without a clear purpose. Your presentation will be judged on content and delivery, not on showing off your ability to insert as many animation gadgets as possible.
- Depend on pointing or waving at the screen to highlight specific material, unless you have a device such as a pointer to focus the audience's attention. Hand waving vaguely at something from your position is of little help to someone viewing the screen from a different angle.
- Have a laser pointer beaming toward the screen during the entire presentation—use it judiciously when you want to draw attention to something specific.

There are other **visual aids** that could be more appropriate than PowerPoint in certain circumstances (see Box 10.5). Indeed, despite its many features and advantages, PowerPoint has one significant drawback. Compared to a whiteboard or flipchart, it is cumbersome for interactive presentations in which (for example) you will be recording ideas offered by the audience during a brainstorming activity. Consider the potential benefits of using an alternative to PowerPoint for a presentation, particularly if your audience will be small.

Software programs are powerful tools for creating visual aids to support an oral presentation for all levels of speakers (Beins and Beins, 2008). Showing images and video clips, displaying unfamiliar words, and listing key points can increase the effectiveness of your presentation by reducing the amount you have to say to present your material. However, in addition to the list of pointers in Box 10.4, keep the following in mind as you prepare a presentation:

Box 10.5 ◉ Visual aids: Alternatives to PowerPoint

There are a number of visual aids other than PowerPoint (or similar presentation software programs) that you may consider using (Gurak, 2000). Depending on the availability of equipment at the presentation venue, you may be forced to use one of these alternatives or you may choose to do so because of the advantages they offer:

- *Whiteboard or blackboard:* Much more versatile than PowerPoint for an interactive presentation, where you want to write down points that arise during the presentation. But consider carefully the time it will take to write on the board, particularly if you want to use diagrams. Your handwriting must be clear.
- *Flipchart:* Similar to a whiteboard or blackboard except it can be moved to different places in the room rather than being fixed in place.
- *Real object or model:* If small, it will be suitable only for very small audiences who will be able to see detail clearly; but if larger, be sure it can be transported. It may have the advantage of giving the audience something tangible to view in comparison to abstract talk or a projected image.
- *Overhead projector:* Before digital projection became the norm, oral presentations were typically supported visually by materials shown on an overhead projector. These devices are now infrequently used.

- *Do not forget who is presenting.* The software program does not make the presentation; it is just an aid to help you make a better presentation. You must not lose sight of the fact that a presentation can be ruined if you are poorly prepared to actually speak. Avoid putting all of your effort into creating a show full of fancy animations and graphics that will distract the audience from concentrating on your central message (Anholt, 2006, p. 94). Have sufficient command of your material that the slides help to make your points without marginalizing the presentation's spoken component.
- *Beware of dividing the audience's attention.* The audience has two things to keep track of when you show information on a projection screen (or use any visual aid, in fact): your voice and the visual display. Use the software presentation to complement what you will be saying, not to present something significantly different. If the screen shows one thing and you speak about something else, audience members' attention may be divided or they may get confused. Worse, they may end up paying attention only to the screen and not listening to what you are saying.
- *Minimize the amount of text.* Most of your presentation should be delivered verbally. Text on the screen should appear in point form with the purpose

of helping the audience follow the progression of your points. Asking the audience to read passages of small-size text while simultaneously listening to what you are saying is a recipe for disaster.

- *Keep the appearance of slides simple.* A cluttered or complicated slide will not achieve the objective of helping the audience grasp the essential message that you want to convey.
- *Include a title slide.* This might show your name, the title of your presentation, your contact details, and perhaps a brief outline of the presentation to follow.
- *Make no more than four or five statements on each screen.* Make each statement in as few words as possible (say, about six words each).
- *Produce large and boldly drawn diagrams.* The audience may not have time to understand a diagram if it is excessively detailed.
- *Information shown on screen should be easy to read.* Use 18- or 24-point text in a clear typeface on slides. Use upper and lower case because IT IS MUCH EASIER TO READ THAN BLOCK LETTERS. Ensure that the text colour is legible against the background—some colour or shade combinations are not easily read.

Visual aids must be large, clear, and in highly visible colours.

- *Use illustrative material instead of tables.* Graphical depictions of information are usually more effective because they are easier to understand than tables. However, tables can be useful if their construction is simple.
- *Consider inserting cartoons.* Well-chosen cartoons can convey a message very effectively and simultaneously help to lighten the atmosphere.
- *Avoid pasting in graphs or tables directly from a written paper.* These often contain more information and detail than can be comprehended readily. Redraw graphs and redesign tables to support the particular message you wish to convey.
- *Be neat.* Sloppily produced work suggests a lack of care, knowledge, and interest.
- *Plan to leave a good impression.* Consider preparing an attractive and relevant final image that can be left on the screen when you have finished speaking and are answering questions. It is sometimes useful to restate the title of your presentation and your name in this final slide.

Rehearse

Few people, if any, are naturally gifted public speakers. It is a skill that must be developed through experience. People new to public speaking often speak far too quickly or too slowly; manage their presentation time poorly by belabouring minor points; fail to engage with their audience (through eye contact, for example); or do not know how to operate the audiovisual equipment. You can overcome such problems through practice and with useful feedback from other students or family members.

With public speaking, practice makes perfect.

Rehearse until there is almost no need to consult your prepared notes for guidance—about 10 times ought to do it. The intent is not to commit the presentation to memory. Rather, rehearsing helps to ensure that you have all the points in the right order and that you have a crystal-clear sense of the key message and path of your presentation. This is vital to success. Rehearsing also enables you to practise the pacing and timing of your presentation so you don't, for example, spend 90 per cent of your allotted time discussing 20 per cent of the planned content of your presentation! It may allow you to consider whether you sound boring or arrogant. It also allows you to manage your spoken material, handouts, and audiovisual resources within the time available (Hay et al., 2005, p. 166).

- *Speak your material out loud!* Not only may you begin to hear problems of logic in your presentation, but you will also have the opportunity to practise pronouncing some of the specialized terms that you might have to employ.
- *Time rehearsals.* Most novice speakers are stunned to find out how much longer their presentation takes to deliver than they had expected, than it took to read quietly, or than it seemed to have taken while they were speaking. Match the time available for the presentation with the amount of material. Build in a few extra minutes to allow for impromptu comments, pauses to gather your thoughts, or the breathtaking realization that the audience is not following your tale! If your presentation is too long, decide what material can be removed without affecting the clarity of the main points. It is better to do the pruning beforehand than to be forced to end your presentation prematurely because you have reached the time limit, or to make revisions in midstream. Only the most confident speakers can think clearly enough on their feet under pressure in front of an expectant audience without preparation, (i.e., to **ad lib**), and still deliver a coherent and effective presentation.
- *Plan the use of your visual aids carefully.* Visual aids can consume time rapidly, for example, if a video clip takes time to start or there is a slow connection when you are loading an external website. In a short

presentation such delays can eat up a disproportionate amount of your time. Consequently, pay careful attention to the use of time in the delivery of multimedia presentations.

- *If possible, record one or two rehearsals.* It is often useful to make a video recording of a trial presentation. Video cameras do not "pull punches" in the same way that an audience of friends and family, sensitive to your feelings, might. If video is unavailable to you, try an audio recording. Find out if you can enliven your presentation through variations in pitch, tone, and pace, or if you use those annoying saboteurs of a good oral presentation: "umm," "err," "you know," "like," and "ahhh."

Rehearse your presentation, including all the prompts and devices that you plan to employ.

Final points of preparation

- *Be dressed and groomed for the occasion.* Although the audience's attention should be on the intellectual merits of your argument, your appearance may affect some people's impression of what you have to say. Looking dishevelled suggests that you do not have respect for your audience or for yourself. It may not be necessary to go as far as Reece's (1999) suggestion of wearing business attire in all situations, but you should look as if you have taken some care about your appearance. Think about what level of attire would be comfortable to wear as well as appropriate for the audience and event where you will be presenting (Anholt, 2006, p. 20).
- *Take water.* No matter how well prepared you are, there is a possibility that you will suffer from a dry mouth which will make it hard to speak clearly. Drinking water during the presentation seems to help!
- *Be familiar with the equipment you will be using.* This includes software programs if you are using a computer. The presentation computer might not be set up exactly the same way as the one you used in practice. Do not be so unprepared that you must exasperate your audience by asking: "How do I switch on this machine?" or "Does anyone know where to find the video player program?" You should have checked these things before your presentation.
- *Be mentally prepared for the possibility of technical problems.* Equipment glitches can throw off some people and make it difficult to pick up the presentation after the problems are fixed. If there is a problem with the equipment, ask for assistance and try to relax. Look over your notes while

When it is your turn to speak, take control of technology, timekeeping, volume, and vision.

you wait so that you keep your mind on the presentation.

- *Make sure the audience can see both you and your visual aids.* Before your presentation, sit in a few strategically placed chairs around the room to see whether the audience will be able to see you and your visual aids clearly. Consider where you will stand while speaking, and take care to avoid obstructing the audience's view of the screen.
- *Check to see if there is a clock in the room.* If not, make sure you can see your watch or have some other way of keeping track of the time.
- *Make sure that everything else is ready.* Are summaries, if you are using them, ready for distribution? Are note cards in order?
- *Make absolutely clear in your mind the central message you wish to convey.* This is critical to a good presentation. Knowing your message will give you the confidence your audience will need to see if they are to have faith in what you are telling them. It also means that if for some unforeseen reason something goes wrong, or you "stall" and lose the plot momentarily, you can take a breath, reflect on your central message, and resume your presentation with the minimum of fuss. It is important, too, because if you do not have your presentation's message firmly established in your own mind, you are unlikely to be able to let anyone else know what that message is.

How to Deliver a Successful Oral Presentation

People in your audience *want you to do well*. They want to listen to you giving a good presentation, and they will be supportive and grateful if you are well prepared, even if you do stumble in your presentation or blush and stammer. The guidelines outlined here are targets you can aim at. No one expects you to give a flawless presentation; however, you can help the audience maintain interest by avoiding the pitfalls noted in Box 10.6.

Audiences support speakers who try to do well.

It will make your presentation more convincing and credible if you remember that the audience is made up of *individuals*, each of whom is listening

Box 10.6 ◉ Directions for lulling an audience to sleep

Turn down the lights; close the curtains; display a crowded slide and leave it in place; stand still; read your paper without looking up; read steadily with no marked changes in cadence; show no pictures; use grandiloquent words and long sentences.

(Booth, 1993, p. 42)

to you. You are not speaking to some single, amorphous body. Imagine that you are telling your story to one or two people and not to a larger group. If you are able to perform this difficult task, you will find that voice inflection, facial expressions, and other elements important to an effective delivery will fall into place.

- *Be confident and enthusiastic.* One of the keys to a successful presentation is your enthusiasm. You have a well-researched and well-prepared presentation to deliver. Most audiences are friendly. All you have to do is tell a small group of interested people what you have to say. Try to instill confidence in your abilities and your message. Do not start by apologizing for your presentation. It sets up the audience with a poor expectation of what is to come.
- *Look interested*—or no one else will be!
- *Speak naturally, using simple language and short sentences.* Try to relax, but be aware that a presentation is not a conversation in a bar. Some degree of formality is expected. Do not use colloquial language unless you have a specific reason for doing so.
- *Speak clearly.* Try not to mumble and hesitate. This may suggest to the audience that you do not know your material thoroughly. Remember that slowing down your speaking from your normal pace will help to make your speech clearer.
- *Speak loudly enough.* Be sure that the most distant member of the audience can hear you clearly. A microphone may be necessary in large venues. Use one if it is available because it will allow you to speak at your normal volume.
- *Use visual aids to complement your speaking.* As mentioned previously, avoid dividing the audience's attention between what you are saying and what you are showing.
- *Engage your audience.* Vary your volume, tone of voice, and pace of presentation. Involve the audience through use of the word *you*: e.g., "You may have asked yourself whether I accounted for the possibility that . . ."
- *Use appropriate gestures and movement.* Step out from behind the computer monitor or lectern and move around a little, engaging with different members of your audience.
- *Try to avoid nervous habits.* Be conscious of distracting behaviour, such as jangling money in your pocket, swaying, pacing back and forth nervously, or saying things like "umm" a lot. You may have detected these habits in your rehearsals.
- *Make eye contact with various members of the audience.* Although this may be rather intimidating, eye contact is very important. It also allows you

to gauge audience response. Look at one person for a few seconds, then another, rather than focusing on just one person.

- *Face the entire audience.* Do not talk to the walls, windows, floor, ceiling, or projection screen. It is the audience—the entire audience—with which you are concerned. If you are showing material on a projection screen, you should have a copy of your slides in front of you so that you do not have to turn your back on your audience to look at the screen. Speaking toward the screen makes it much more difficult for people to hear you clearly.

> ***Pay attention to your audience: Can they hear? Can they see? Are they bored? Do they understand?***

- *Pay attention to the audience's reaction.* If the audience does not seem to understand what you are saying, rephrase your point or clarify it with an example.
- *Direct your attention to the less attentive members of the audience.* This may not be as reassuring as focusing your presentation on those whose attention you already have, but it will help you to convey your message to as large a part of the audience as possible.
- *Help the audience to hear what you are saying.* Have key words and unusual words visible on the screen for the audience to read.
- *Stop speaking when a new slide is first shown.* This is to allow the audience time to study the display. Then, take a moment to familiarize your audience with elements of the visual aid, such as axis labels, remembering that unlike you they have never seen the image before. There is no point in telling people what the image means before they have had a chance to work out what it is about (Pechenik, 2007, p. 262).
- *Point to the audience's screen, not yours.* Some speakers new to digital projection will point with their finger to images on a monitor in front of them, believing perhaps that the audience can see what they mean, instead of the actual projected image.
- *Use a pointer judiciously*. Use a laser pointer or the cursor only to draw the audience's attention to a specific part of the screen. Do not flash it about the screen unnecessarily (for example, circling an item for the entire time you are speaking about it). Once you have drawn the audience's attention to something important on the screen, further pointer movements are distracting. And be aware that laser pointers can significantly accentuate a slightly trembling hand, making it much more noticeable (Pechenik, 2007, p. 263).
- *When you have finished with an illustration, remove it.* The audience's attention will be directed back to you (where it belongs) and audience members

will not be distracted. Do not speak about a topic that is different than the one on your visual display.

- *Keep to your time limit.* Audiences do not like being delayed, but you should take care not to rush at the end. Last-minute haste may leave the audience with a poor impression of your presentation. Watch the time as you proceed. If it is apparent that you will run out of time, let the audience know the remaining topics you planned to cover and then jump straight to the conclusion. This is likely to be the most effective way of summarizing the rest of your presentation within the time limit. Of course, if you have rehearsed properly, this problem should rarely occur!

How to Cope with Questions from the Audience

The post-presentation discussion period that often follows an oral presentation allows the audience to ask questions and to offer points of criticism to which the presenter may respond (Alley, 2003, p. 3). It is an important part of the overall presentation that can completely change an audience's response to you and your work. Anticipate the requirement to field questions instead of expecting to be able to depart immediately after you finish speaking. Take care to be thorough and courteous in your response to comments and questions.

> ***Anticipate that you will be expected to answer questions following your presentation.***

- *Let the audience know whether you will accept questions in the course of the presentation or after the presentation is completed.* Usually audiences wait until the end before asking questions. People tend not to like interrupting a speaker's flow unless the speaker has specifically invited questions during the presentation. In considering whether to do so, be aware that answering questions may disturb the flow of the presentation or may anticipate points to be addressed at some later stage within the presentation.
- *Stay at or near the lectern throughout the question period.* This is still a formal part of the presentation. Do not return to your place in the audience until it is clear the question period has concluded.
- *Be in control of the question period.* However, if there is a chairperson, it is his or her responsibility to keep track of the time and sometimes to keep track of the people who want to ask questions.
- *Address the entire audience, not just the person who asked the question.* Although one person asks a question, your answer needs to be given to everyone in attendance.

- *Repeat a question if it is difficult to hear*. This ensures that you heard the question correctly. Repetition will also help audience members who may not have heard the question clearly.
- *Ask for clarification of any questions you do not understand*. Do not proceed with an answer until you have a clear understanding of what is being asked.
- *Take questions from all parts of the audience*. Take care to receive inquiries from different people before returning to any member of the audience who wants to ask a second question.
- *Search the whole audience for questions*. Scan the entire room, not just the area closest to you.
- *Always be succinct and polite in replies*. You should be courteous even to those who appear to be attacking rather than questioning honestly, for two reasons. First, if you have misinterpreted the intent of the question—perceiving an affront where none was intended—you will avoid embarrassment. Second, one of the best ways of defusing improper criticism is through politeness. Even if there is no doubt that someone is being hostile, keep your cool.

Treat the question period as seriously as the main part of your presentation.

- *Avoid concluding an answer by asking the questioner if you have answered the question satisfactorily*. Argumentative questioners may take this opportunity to steal the limelight, thereby limiting the discussion time available to other members of the audience.
- *Deal with particularly complex questions or those requiring an unusually long answer after the presentation*. If possible, give a brief answer when the question is first asked, and then offer to meet individually with the questioner.
- *If you do not know the answer to a question, say so*. Do not try to bluff your way through a problem, as any errors and inaccuracies may call the content of the rest of the presentation into question.
- *Difficult questions may be answered by making use of the abilities of the audience*. For example, an inquiry might require more knowledge in a particular field than you possess. Rather than admit defeat, it is sometimes possible to seek out the known expertise of a specific member of the audience. However, it is your presentation, and so deflecting questions to others should be done sparingly and should not be used as a convenient way to avoid answering a question yourself if you have the knowledge to do so.
- *Smile*. It's over!

Further Reading

Alley, M. (2003). *The craft of scientific presentations*. New York: Springer Science+Business Media.

- Distributed throughout the book are discussions of 10 critical errors that presenters often make. Learning from the mistakes of others is an excellent way of preparing for your presentation and avoiding making those mistakes yourself. Advice for handling questions, "even the tough ones" (p. xiv), is included.

Anholt, R.R. (2006). *Dazzle 'em with style: The art of oral scientific presentation* (2nd ed.). Burlington, MA: Elsevier.

- This is an informative guide filled with useful suggestions for preparing and presenting oral presentations. One of the four chapters is devoted to various aspects of effective delivery.

Booth, V. (1993). *Communicating in science: Writing a scientific paper and speaking at scientific meetings* (2nd ed.). Cambridge, UK: Cambridge University Press.

- Chapter 2 is an entertaining and informative exhortation to speak well at professional gatherings.

Bowman, L. (2001). *High impact presentations: The most effective way to communicate with virtually any audience anywhere*. London: Bene Factum.

- Encourages speakers to use a conversational style to make a more impressive and effective presentation.

Calnan, J., and Barabas, A. (1972). *Speaking at medical meetings: A practical guide*. London: Heinemann.

- This is a marvellous little book. Although some of the material applies only to the medical profession, most of the content of this easily read, humorous volume is applicable to speakers in other fields.

Carle, G. (1992). Handling a hostile audience with your eyes. In D.F. Beer (ed.), *Writing and speaking in the technology professions: A practical guide*. New York: ieee, 229–231.

- Carle's paper outlines an interesting three-step strategy for dealing with hostile audiences.

Eisenberg, A. (1992). *Effective technical communication* (2nd ed.). New York: McGraw-Hill.

- Chapter 15 consists of a long discussion on preparing and giving an oral presentation. Some useful notes on body language and on writing out a presentation are included.

Gurak, L. (2000). *Oral presentations for technical communication*. Boston: Allyn and Bacon.

- This book is a comprehensive guide to oral presentations. Part 1 contains chapters on dealing with nervousness and on techniques to build confidence in speaking. Part 3 describes four types of presentations with different objectives; Part 4 considers the nature of different audiences, including non-experts.

Hay, I., Dunn, K., and Street, A. (2005). Making the most of your conference journey. *Journal of Geography in Higher Education, 29*(1), 159–171.

- This article extends this chapter's discussion to look further at effective delivery and at ways of taking advantage of a conference experience to forge valuable professional relationships.

Jenkins, A., and Pepper, D. (1988). Enhancing students' employability and self-expression: How to teach oral and groupwork skills in geography. *Journal of Geography in Higher Education, 12*(1), 67–83.

- An extensive discussion of the academic and vocational rationale for teaching and learning oral communication is included in this article.

Jones, M., and Parker, D. (1989). Research report development of student verbal skills: The use of the student-led seminar. *The Vocational Aspect of Education, 41*(108), 15–19.

- In its introduction this article outlines some of the vocational reasons for learning to speak effectively.

Montgomery, S.L. (2003). *The Chicago guide to communicating science.* Chicago: University of Chicago Press.

- Chapter 13 provides a helpful guide to scientists on giving high-quality presentations to both professional and lay audiences.

Street, A., Hay, I., and Sefton, A. (2005). Giving talks in class. In J. Higgs, A. Sefton, A. Street, L. McAllister, and I. Hay (eds.), *Communicating in the health and social sciences.* Melbourne, Australia: Oxford University Press, 176–183.

- This reviews the advantages and disadvantages of various visual aids and suggests that alternatives to PowerPoint may be more suitable visual aids in some circumstances.

11 Writing for the Media and Public Audiences

How is the world ruled and how do wars start?
Diplomats tell lies to journalists and then believe what they read.

—Karl Kraus

Never before in history has it been possible for any individual, group, or community to self-publish worldwide on equal par with the world's greatest governments, universities, and corporations.

—Lone Eagle Consulting

Key Topics

- Communicating with public audiences through the media
- Using media releases to disseminate information
- Disseminating information by self-publishing on the Internet

Geographers and environmental scientists are often interested in ensuring that the general public know of their work and their views on important issues. This chapter offers some guidance on preparing material for communicating to the public through the journalistic media and through self-publishing on the Internet. While advances in technology have radically changed what is meant by "the media," there remain some basic principles to follow when writing for public audiences. Effective writing and presentation are key to making your efforts worthwhile. Following up on a media release is recommended, and some comments on participating in a successful interview are offered.

Various forms of media provide access to vast, valuable, and powerful audiences.

Communicating with Public Audiences through the Media

Previous chapters have discussed effective communication with what are often small audiences. When you write for your courses as an undergraduate, your audience may ultimately be as small as one—the marker. But later you may pursue a career that requires you to write essays or reports, or deliver talks, that will be received by larger groups. This chapter is devoted to communicating through the media to a mass audience—the public. Box 11.1 lists the main reasons why geographers and environmental scientists need to communicate with the public.

Communicating with public audiences through the media is something that you are more likely to encounter in a career situation. Being asked to write a media release is not a common exercise in undergraduate courses. This book is oriented primarily to the common undergraduate experience; nevertheless, this topic fits with the objective of discussing and developing better communication skills. Some professors do indeed ask their classes to practise public communication by writing sample media releases in order to help develop skills for post-graduation situations. Meanwhile, as an undergraduate you may be involved in extracurricular activities that require, or would benefit from, communicating with the public. Being exposed to the idea of communicating through the media, you may even see opportunities to disseminate a message that you may not have realized existed.

Box 11.1 ◉ Why communicate with a public audience?

There is a wide variety of reasons for communicating with the public, as follows:

- To inform and raise awareness among the general population about geographical and environmental issues;
- To communicate the results of your scholarly activities to broad and interested audiences;
- To offer your selection of facts and views on a particular situation or issue
- To help establish a favourable public profile for your work, your discipline, and the organization you represent, which may help you find sponsors and other forms of support, including financial;
- To promote and secure support for activities you are involved in (e.g., presentations, public meetings, research in progress).

Types of media

Before the Internet became a dominant presence in the 1990s, communicating with a mass audience was usually dependent upon sending written information to a journalism-based media outlet, such as a newspaper, magazine, television station, or radio station. If the story interested a reporter or editor, it might have been selected for publication or broadcast. Between submission and dissemination, a piece would be filtered by the reporter or editor, only parts of it might be used, and its creator's intended meaning might be lost. The Internet has profoundly changed the media industry. Let us consider the choices now available to someone who wishes to disseminate information to mass audiences.

Journalism-based or "filtered" media

Communicating your story to the public may still mean preparing material in the form of a **media release** and sending it to a journalism-based media organization, where you can still expect that filtering by a reporter or editor will occur. A media release can be considered a traditional approach to disseminating information for public consumption, although the media themselves are not limited to the traditional methods of dissemination. Many media organizations now have multiple outlets for their information, including websites to complement (or replace) traditional print materials or broadcasts.

Self-publishing on the Internet

The Internet offers intriguing options for **self-publishing** that can spare you the uncertainty of depending on other people to decide that your story will be of interest to their audience. The freedom of self-publishing must be balanced, however, by uncertainty over whether your message will reach your intended audience and whether you will be taken seriously. The principles that underlie writing a release to be submitted to a media outlet are also relevant to material that will be self-published. Options and considerations for self-publishing will be discussed later in the chapter.

A continuum of media options

A range of media outlets exists, with varying degrees of credibility, quality, and connection with the public. The dominance of traditional journalism subject to strong editorial control has been diminished with the dramatic rise of the web as a means of communicating information more broadly. The many long-standing, highly respected media organizations have been joined by new media channels that exist only on the web. Blogs, for example, are subject to minimal editorial control and often have biased agendas. A blog may in fact represent the views of a very small number of people, even just one person.

The use of such outlets to publish your material differs from self-publishing for, as with journalism-based media outlets, your story may be trimmed or altered before it reaches the public.

If you want to convey information to a mass audience, you should consider all options for reaching your target audience. However, in Table 2.2, you were encouraged to evaluate the credibility of web pages as sources of information. You should give similar consideration to whether you would want to have your information presented by or associated with a particular website or organization. Do you want to risk sending your material voluntarily to people who might distort stories to fit their agenda, who could make a mistake when citing your material to prepare their version of a story, or who create a poor online presentation?

The rest of this chapter will discuss two approaches to communicating with public audiences. Attention will be paid first to preparing an effective media release for communicating with the public through a journalism-based media outlet. Then self-publishing on the Internet will be discussed. In fact, however, a clear line cannot be drawn to separate the two approaches since a continuum can be seen to exist between journalism-based media and self-publishing.

Using Media Releases to Disseminate Information

Choosing to communicate with the public through a journalism-based media outlet can provide exposure and legitimacy that may be difficult or impossible to obtain with self-publishing. The challenge is to provide the media outlet with material that it will pick up rather than discard.

The most commonly employed means of communicating with the media is through a *media release* (also sometimes known as a *press release*). Media releases are a major source of the information used by journalists. They generally set out the core information about something the writer hopes will get favourable coverage. A media release can be important for helping reporters to get their facts right.

Media releases were formerly known more commonly as news releases or press releases. Carney (2008) criticizes the use of "press release," however; he states that "'press' refers to print media, and broadcast media have been known to discard 'press releases' as an indication that the sender is clueless about media relations" (p. 102). "Media release" is a broader term.

> ***A media release is a vital bridge between your academic and professional work and the mass media.***

A media release is often your first contact with the media about any story. And it may, in fact, be your only contact.

You should try to ensure that the release is well-written, comprehensive, and captivating (see Box 11.2). Try to provoke interest and set the tone for the ways in which the story might be covered subsequently. Targeting a media release is also an important task that is likely to influence whether or not it is successful (see Box 11.3).

In his provocatively named piece "Die! Press Release! Die! Die! Die!" Tom Foremski (2006) suggests that the traditional approach to writing press or media releases should change. Foremski is a former *Financial Times* writer who resigned to start a blog called Silicon Valley Watcher. He finds the traditional structure of media releases too formulaic and the typical content "nearly useless" for a journalist.

While you may write a release as a coherent piece with your "spin" or perspective on a matter, Foremski suggests that your spin is likely to be replaced by the journalist's. He feels that users are really looking for basic factual material to use in their writing. Instead, he advocates that you: "Deconstruct the press

Box 11.2 What are users of media releases looking for?

Remember, mass media are essentially profit-driven businesses. The staff at newspapers, television or radio stations, and websites aim to sell or distribute as much of their information product (and associated advertising) to as many readers, viewers, or listeners as possible. For this reason, they are most likely to be interested in stories with the following features:

- The release provides information that is new or up-to-date (e.g., an economic geographer's explanation for a recent rise or fall in the national currency).
- The information is unusual or unexpected (e.g., discovery of a new species of cold-water coral on the continental slope).
- The story is captivating (e.g., research showing that a particular species has disappeared from a certain area because of habitat reduction or fragmentation caused by clear-cut logging).
- The information is relevant to the geographic range of the medium's audience. Remember that most media outlets have specific markets to which they aim to make their stories interesting and relevant. Thus, a small-town newspaper may carry stories on environmental issues related to the surrounding region that a larger regional or national organization will not be interested in covering.
- The news item is relevant to the interests of a large part of the medium's audience. Relevance of readers' interests may trump relevance of geographic range: consider that many stories are specific to one part of a country, but there may be broader interest in the story.

Box 11.3 ◉ Targeting a media release

To help make the best choice of media outlet for your release, ask yourself the following questions:

- What relevant outlets exist?
- Are there particular outlets or presentation modes that are better suited than others to your material (e.g., a story for which you have an intriguing video needs to be sent to an outlet that can show the video)?
- To what audiences is the content relevant (e.g., local, provincial, national, international; conservative, wealthy, environmentally active)?
- For print media, how often do these publications appear (daily, monthly, quarterly, annually)?
- How much space or time is typically devoted to specific issues?

release into special sections and tag the information so that as a publisher, I can pre-assemble some of the news story and make the information useful" (Foremski, 2006, unpaginated). These sections should include basic description, lists of quotations from relevant parties, factual information, and links to relevant material (both within the previous material and additionally in a separate section).

Large organizations, such as universities and companies, tend to have staff who specialize in public relations. If you are representing the organization in your media release (e.g., using the organization's letterhead), you may be required to have your release approved before it is disseminated or to work with the staff during its development. Even if this is not a requirement, you might still consider seeking the advice of trained people to improve your release.

Selecting the right media outlet is vital to the success of your story.

Check the regular features pages or sections of the publications or websites you are targeting. Many publications have sections devoted to particular topics, such as environment, real estate, or higher education. Some sections may appear only periodically. Not only might this be of significance in terms of the timing of your media release, but it will also point out the sorts of broad issues that the outlet considers most important.

How to write a media release

Breckenridge (2008) examines how technological developments have expanded the capabilities of those preparing media releases. While some advocate the continued use of the traditional inverted-pyramid style, i.e., text-based releases where the importance of information decreases from the top down, others suggest abandoning that limited format (see discussion above of Foremski, 2006). In its place, they favour a more direct approach to providing information and the adoption of more varied options that reflect the new digital communication environment. The two different approaches are discussed in more detail below.

Information common to all media releases

Certain principles apply to the writing of any media release:

- Format the media release to be highly legible.
- If you are representing an institution or organization, write on its letterhead if possible. If letterhead is not available to you, be sure to give your name, that of the institution or organization with which you are associated, and full address and contact details at the top.
- Mark the document "Media Release" at the top and add a short, engaging title that clearly explains what the release is about.
- Date the media release.

Structure of inverted-pyramid-style media releases

In many respects, a traditional inverted-pyramid-style media release does not follow the principles of good writing, but it is the way in which news stories have traditionally been written. It is certainly different from other forms of written communication with which you might become familiar in higher education. The inverted-pyramid style does, however, require attention to structure. The media release is designed to catch the reader's attention (i.e., hook the reader), provide a context, and deliver easily digested quotations. In writing a media release, it is all too tempting to write too much. You do not just want to attract the audience to your story; you also want to keep their attention.

A traditional media release follows an inverted-pyramid structure so that whole paragraphs near the end can be deleted without losing the story's central message.

Here is some advice for structuring an inverted-pyramid-style media release:

- Use a catchy headline and set it in large, boldface letters.
- Make your most important points first; and then provide others in declining order of importance. Typical news stories start with the most important information and then go on to provide explanatory paragraphs with

other details in declining order of importance. There are several reasons for this. First, the reader's attention is captured immediately. Second, because most people do not have time to read an entire newspaper or news website, it is helpful to them if they can gain a broad overview of the news of the day by reading headlines and introductory paragraphs. Third, stories often have to be edited down to fit the available space. With the inverted-pyramid structure, editors can be fairly confident that by removing material from the end of the story, they are not removing any vital information. For this reason, be sure that any release you write will still make sense if paragraphs are removed from the end.

- Keep the opening paragraph ("the lead") brief. Be sure it makes an impact and highlights the newsworthy angle of the story. This forces the writer to focus on the main story hook.
- The opening paragraph provides the "story hook" (Williams, 1994) and should tell the *who*, *what*, *when*, *where*, and *why* of a story.
- The second paragraph can provide a "secondary hook" and should explain in more detail the information from the first paragraph.
- Keep paragraphs short. Try limiting them to a single sentence. Note the clear difference in paragraph construction compared to that normally described for paragraphs (see Box 1.1).
- The release should conclude with "END" or "-30-" to let the reader know what is media release material and what is subsidiary information, like contact details and photographs. (The use of -30- is a traditional practice with historical roots.)
- Conclude the release with the name and full communication details of the primary contact person who can be called, emailed, or texted for more details. A secondary contact can also be supplied. For stories relevant only for a short time, it is essential to give after-hours and weekend contact details as well.

Make yourself available to reporters who are following up on your release.

Evidence and credibility

- Near the top of your story, give a few details about yourself or your organization, such as the function it serves and the number of members.
- Do not assume reporters are familiar with you or the organization with which you are associated. Mention its full name early in the release with the initials in parentheses.

Quotations from prominent figures lend authority to a media release.

- Use quotations wherever possible. These give the story vitality and credibility. Quotations must accurately represent what a person said. You may be proactive and ask someone to provide a quote. If possible, check with the people you quote to make sure you have represented their statements correctly. An adversary may not wish to co-operate with you in this exercise, so responsibility for accuracy would lie entirely with you.
- Attribute statements to a particular person and not to some anonymous "spokesperson." Give the person's full title, name, and position.
- Give each person's full name, and be sure to include some phrase that makes it clear to readers what that person's claim to authority in this story is.
- Support statements and claims with "hard" evidence, such as statistics, quotations, and references to government documents and other reports.
- Be accurate with your facts.

Language

- Be sure the release is clear, concise, and arresting. Try to keep it to a page or less (about 300–400 words). Keep the sentences short; leave out unnecessary words. Readership can fall off by as much as 80 per cent after the third or fourth paragraph of a news story (Williams, 1994, p. 6). With this in mind, a short and to-the-point release should be the goal.
- Keep language simple and concise (e.g., *now*, not *at this point in time*; *said*, not *exclaimed* or *stated*; *before*, not *prior to*; *regularly*, not *on a regular basis*).
- Avoid jargon and clichés.
- Be sure all spelling is correct and that all names and titles are accurate.
- Unless the story is about something that is yet to happen, use the past tense.
- Use the active voice rather than the passive voice. In the passive voice, the subject receives the action expressed by the verb ("The tree was hit by the moving car"), whereas in the active voice the subject performs the action of the verb ("The moving car hit the tree"). In general, the active voice is more lively and succinct.
- Use positive rather than negative statements.
- Engaging with the media can be tiresome and disheartening. A media release that may have taken many hours for you to write may not even make it beyond the recycling bin. If the story does reach the public arena, it may look nothing like your original. Walters et al. (1994) reported that, on average, press releases were cut in half by reporters and editors, from an average of 435 words to 210; from 13.5 paragraphs to 6.3; and from 18.4 sentences to 9.3. The reading level of releases was lowered ("dumbed down") from a Flesch-Kincaid grade level of 15.0 to 13.6, and

the average length of words reduced from 5.3 to 5.1 letters. In such cases the reporter and editor are making your story more palatable and hence more digestible for the audience. Do not regard the changes as a slight against your professionalism and integrity. It is simply that the journalist and editor know the right level of language for their readers.

Photographs and supplementary information

- It can be very helpful to submit a good-quality, relevant photograph (or other illustrative material, such as a map) with your story, particularly if you are submitting it to a small media outlet with limited resources. Photographs should be of high quality and in a standard format, such as JPG, TIF, or PNG. Though larger organizations may not use your photo, it may give their staff some ideas for pictures of their own. Include a typed **caption** for the photograph (be sure the caption answers *what*, *where*, *when*, and *who* questions).
- In some cases a picture can make a story. For instance, a story supplemented by a recent close-up photograph, taken in Nova Scotia, of an eastern cougar, a species that was last seen in the Maritimes in 1938, might make the front page of most Canadian national newspapers. An unillustrated report of its sighting is less likely to figure prominently.

> ***Consider including a high-quality, relevant photo with your media release.***

- Try to avoid large-group photos. When reduced in size for publication these tend to lose relevant details.
- If you find that the complexity of the story warrants it, consider including with your release a separate "fact sheet" that explains complicated matters clearly. If your release describes a paper or research report you have written, it may be helpful to send a copy along with your media release.

Timing

- For a story that publicizes an upcoming event, give the journalist plenty of time to react to your media release (several days or weeks if possible). A media organization's schedule and priorities are unlikely to match yours. If you allow more time, there is a better chance your story will be used.

How to format a media release

Although not all media releases will follow it exactly, the basic outline for a media release is as shown in Box 11.4.

Box 11.4 ◉ Suggested layout of a media release

Letterhead
FOR IMMEDIATE RELEASE [or Embargoed until time, date]
Date
Headline

Introductory paragraph that answers *Who*, *When*, *Where*, *What*, and *Why*?

A second paragraph explaining in more detail the information from the first paragraph.

A third paragraph that includes a quotation attributed to some prominent person.

A fourth paragraph that includes some more information, perhaps another quotation.

Subsequent paragraphs add supplementary information that can be removed by an editor without making the earlier parts of the release incomprehensible.

-END- [or -30-]

For further information, contact:
Full title and name of contact(s)
Phone number(s)
After-hours phone number(s) [if applicable]
Email address(es)
Website address [if relevant]

Box 11.5 gives an example of a media release written to announce a new partnership for commercializing environmental technologies set up at a Canadian university.

Box 11.5 ◉ Example of a media release to make an announcement, or to publicize an event or innovation

Media Release
For Immediate Release
October 19, 2012

Getting Green Faster

Atlantic Canadian researchers developing green technologies will soon be able to get them to market more easily with the October 19 launch of the Atlantic Green Chemistry Innovation program at Saint Mary's University.

"This program will provide the region's brightest minds with specialized expertise and resources that will accelerate the pace of innovation and ensure more discoveries make it to market," said Dr Kevin Vessey, Assistant Vice-President, Research at Saint Mary's University.

Every day, researchers across Canada and around the world are discovering chemical products and processes with the potential to address global challenges of sustainability and climate change. At the same time, demand for cleaner, less energy-intensive chemical alternatives has never been greater.

Lynn Leger, Director of Commercial Development with Green Centre Canada, says the new initiative dovetails with her organization's active commercialization of "smart" chemistry discoveries across Canada for the past four years."

"We have great potential with discoveries from Atlantic Canadian universities and have made a decision to invest in on-the-ground resources as a result of our partnership with Saint Mary's and Springboard Atlantic."

"This is a great example of how Springboard Atlantic and its members can partner with a national centre to bridge the gap between academic labs and the marketplace," says Chris Mathis, President and CEO, Springboard Atlantic.

"This will accelerate commercialization of post-secondary research in this sector. Thanks to the support from ACOA's Atlantic Innovation Fund, Springboard is helping increase economic outcomes from our research."

The Saint Mary's Atlantic Centre for Green Chemistry boasts 23 green chemistry researchers working in nine universities. The Centre's director, Dr Robert Singer, said his members will take full advantage of the new program by leveraging the strengths of each of the partners.

The Partners:

GreenCentre Canada *uses its experience in product, application and business development, intellectual property management, and scale-up manufacturing to help advance green chemistry technologies to market.*

Springboard Atlantic *is a network of technology transfer professionals with a sustainable mandate to accelerate academic innovation.*

Saint Mary's University *established the Atlantic Centre for Green Chemistry in 2009 to enable university chemistry researcher from across the region to collaborate on research projects.*

-30-

For More Information:
Steve Proctor
Communications Manager
Saint Mary's University, Public Affairs
(902) 420.5513
Email: steve.proctor@smu.ca
www.smu.ca/administration/publicaffairs/

Media release used with permission of SMU Public Affairs

How to enhance a media release for the digital environment

The core of a media release is the information that it contains, and its effectiveness lies mostly in how well it is written. However, in a digital world it is possible to add certain elements that will make the release more effective. Chief among these elements are links to websites containing more information. Remember that for a media release to be concise, it can contain only the most important details about the story. But by inserting links, you can direct the reader to further relevant information. Links can be inserted at suitable points in the body of the release or included with the contact details at the end.

Some other elements that could be added are supplementary photographs, video clips, and audio segments. If these elements cannot be included directly in the release itself because of size limitations for emails or because you are faxing the release, provide links to the websites where the materials can be accessed.

However, while some supplementary materials can impress the viewer, usually these elements will not stand on their own. Text will still be required to give the essential information or explanation about the story. Remember that the success of a media release depends on how well you convey the story and whether you are able to convince a reporter or editor of its importance.

How to follow up on a media release

It is impractical to expect an immediate reply to your release unless you have a particularly urgent or engaging story. However, if you do not hear back within a couple of days after submitting it, consider telephoning to see if the story is of any interest and if you need to send more information. This is also a good opportunity to try to convince a reporter that your story is worth covering, but don't make a nuisance of yourself. It is often difficult to foresee whether any particular story will get coverage. Some stories will be reported. Others will not.

> ***Before you do a media interview, get clear in your mind the key messages you want to convey.***

After you have sent out a media release, you may be contacted by a reporter seeking follow-up information. During the interview take time to explain your points, but try not to ramble or dwell on marginal issues. If you have discussed an issue that is critical to the story, ask to have the reporter's notes read back to you to ensure that he or she has "got it right." You are unlikely to encounter an objection, as this will help him or her avoid disputes about material after the information has been disseminated.

Disseminating Information by Self-Publishing on the Internet

Self-publishing on the Internet includes options such as building your own website, posting videos or photographs online (e.g., on YouTube or Instagram), writing a blog, and using social networking sites like Facebook or Twitter. In other words, you can join the growing ranks of *citizen journalists*, people without formal journalistic credentials who use the Internet to disseminate opinions and information on their own or in collaboration with others (Glaser, 2006). The principal advantage of self-publishing is that you can be the sole mediator of the information that the public reads. You are in control. Unless you engage someone to read your work before you send it out to the Internet, no reporter will modify your material or make mistakes in writing the story, and no editor will demand changes or cut out parts that you consider important.

Opportunity comes with responsibility

The absence of mediators may persuade you to become a citizen journalist, but it does not absolve you of ensuring your work meets a high standard. If you want to self-publish, are you prepared to take on the editorial responsibilities? An editor can point out errors, omissions, and parts of your writing that need improved clarity. Recall that in Chapter 1 you were advised to have someone read your work and offer comments before submitting it. Should it be any different when you put your work on the Internet?

> ***When self-publishing, you must take on editorial responsibility for the material.***

Consider the potentially massive audience on the Internet (theoretically, anybody who has online access). Some in that broad audience will assess your credibility and the quality of your work with critical evaluation techniques, just as you are encouraged to do in Table 2.2 for websites you consider using as sources.

Is self-publishing effective?

Websites, blogs, Facebook, Twitter, YouTube: these are some of the current Internet applications that allow individuals to disseminate information to mass audiences. As the second quotation at the beginning of this chapter states, individuals now have access to some of the same communication portals as major institutions do. Whereas with traditional media outlets the challenge is to make your story interesting enough to be selected for publication, with self-publishing the challenge is to ensure that your material will reach the target audience.

Ask yourself whether self-publishing will achieve your objective of communicating with a mass audience.

There are many examples of online material "going viral," with audiences growing exponentially within a short time as people forward the material to their friends and others learn about the phenomenon through the media or on websites. In most cases, the explosion of interest was not foreseen, and the material could just as easily have remained in obscurity. If you are considering self-publishing, you must be prepared to accept that putting your material online is no guarantee that people are going to see it. When considering self-publishing options that require more effort, such as creating an effective website, it might be more worthwhile to focus that effort on working with the journalistic media if your objective is to reach broader audiences.

With self-publishing, your audience may be small but you might decide that reaching that audience will achieve your objective and you will still retain editorial control. For example, if you want to promote an upcoming event at your institution, is it a problem that your information is not distributed to the furthest nooks and crannies of the Internet? Using Facebook and Twitter to communicate with fellow students at your institution or people in the local community might be a better way of getting your message to the people that matter.

Sending unsolicited emails is not an acceptable way of disseminating information to mass audiences.

Email and the web were the two most important elements of the early Internet for most people. But email is not considered suitable for disseminating information to mass audiences. Some are now suggesting that email will become passé, an outmoded form of communication replaced by other technologies. Two reasons are the scourges of **spam** and computer viruses. Some mass spammers have been prosecuted although this area of law is new and what constitutes illegal activity is still being clarified. To prevent infecting their computers, email users are encouraged not to open file attachments from unknown senders because that is how viruses are commonly transmitted.

However, it is acceptable to send bulk email messages by means of **listservers**. For example, the Canadian Association of Geographers has a listserver (CAGList). In contrast to spam, a message sent through a listserver is not unsolicited, since the recipients will have signed up to receive messages sent through it. Even if you do not know all the people in the group personally, their decision to join means that they have agreed to accept messages from you or anyone else on the list. Messages from all senders are sent to all list members. However, listservers should only be used to send messages of topical relevance, not for general promotion. Not all recipients will be equally interested in all messages. If you are not certain that a segment of the listserver population will be interested, you should not send the message since those who are not interested will consider it to be spam. Angering the majority of recipients is not a formula for successful dissemination of your message!

Content and presentation

To place information on the Internet, consider three options: (1) building a website, (2) writing a blog, and (3) producing a video. Many of the same principles that were described previously for writing a media release are applicable to these self-publishing options. Consider the characteristics of your target audience and tailor your material to that audience's level and interests. Write in a style that catches people's interest. Remember to place more important information near the beginning because the reader's interest will drop off toward the end. If you have control over appearance, use font types and sizes that permit text to be read onscreen without straining. Instead of single, wide columns, use several columns. Ask yourself whether the colour of the text is legible against the colour of the background, particularly if the background is an image. Carney (2008, p. 143) notes that many self-made websites do not provide important information, such as contact details. Be sure to check that simple details are included and accurate because the website is representing you and your organization to the world.

Most blogs are developed with a program offering a limited selection of design choices and although you have some control over appearance, you will have more if you build a complete website on your own. Building a simple but effective website is not difficult if you have some programming experience or good graphic design skills. It is strongly recommended that you consult a book or online resources that describe the principles and methods of website design. Over time, your site or blog may attract a large enough audience to make your efforts worthwhile.

Producing a video opens up a new area of potential in communicating to an audience. Again, many of the basic principles of communication described for other avenues apply: make sure the principal message is stated clearly and supported with evidence; place clarity and content above empty visual effects; and consider the structure of your presentation from the perspective of the viewer. At the same time video is a much different form of presenting information, so think of ways to take advantage of its unique characteristics:

for example, to stir emotions through visual imagery, to show movement and action, and to use sound for effect.

Conclusion

The end results of writing for public audiences can be enlightening, both to you and to your readers. At worst you will have explored an alternative genre of writing. At best you will have achieved your bit of fame and sparked interest in, and knowledge about, your activities as a geographer or environmental scientist within the wider community. The various media are vehicles by which we can endeavour to communicate the results of our labours to audiences larger than a group of markers, colleagues, families, and friends. If we want larger audiences within the general public to value our work and other contributions to the community, it is essential that we develop and refine our skills in presenting information to and through the media. Self-publishing on the Internet is an emerging option, although you must consider whether material distributed online will actually reach the target audience. Putting your material on the Internet does not necessarily mean people will see it. In fact, if you have a particularly engaging story to share with the general public, you may find that disseminating the information through the journalistic media, despite its drawbacks, may be more effective.

Further Reading

Baines, P., Egan, J., and Jefkins, F. (2004). *Public relations: Contemporary issues and techniques*. Amsterdam: Elsevier Butterworth Heinemann.

- The book provides an overview of public relations that is useful as a general introduction to the field.

Bland, M., Theaker, A., and Wragg, D. (2005). *Effective media relations: How to get results* (3rd ed.). London: Kogan Page.

- Written for the British audience, this book is clearly laid out and informative. Several chapters discuss ways of "handling the media," such as conducting successful interviews and talking to the press.

Bonner, A. (2004). *Media relations*. Edmonton: Sextant.

- The author emphasizes that the writers of media releases "need to be clear in their own minds of what they want to say before they start talking to reporters" (p. 20). Specific strategies for meeting this objective are presented.

Breckenridge, D. (2008). PR *2.0: New media, new tools, new audiences*. Upper Saddle River, NJ: FT Press.

- "2.0" refers to the idea that the field of public relations has changed completely with the introduction of the Internet. It covers new options for communicating, such as social media and social networking.

Carney, W.W. (2008). *In the news: The practice of media relations in Canada* (2nd ed.). Edmonton: University of Alberta Press.

- This is a useful guide for understanding the media and media relations in Canada. Tips are provided for participating in successful interviews.

Mondo Code. (2006). *Mondo Times.* www.mondotimes.com/index.html (accessed 20 November 2013).

- This website is a worldwide, searchable guide to print, audio, and visual media.

Telg, R. (2000). *Developing Effective Media Relations for Your County Program.* http://edis.ifas.ufl.edu/WC020 (accessed 6 October 2009).

- This online article provides a short explanation of how to understand the news media and how to develop a media relations strategy.

12 Succeeding in Examinations

Examinations are formidable even to the best prepared, for the greatest fool may ask more than the wisest man can answer.

—Charles Caleb Colton

Key Topics

- Purposes of examinations
- Types of examination
- How to prepare for an examination
- Techniques for passing different types of examination

Some people flourish in exams, doing their best work under conditions that others might find highly stressful. In part, good performance may be a consequence of a person's particular response to stress, but it is more likely the result of good exam technique. This chapter discusses exams, their types, and strategies for success.

The term *exam* is used here to refer to end-of-term examinations that are usually two to three hours in duration. With some adaptation, the information in this chapter also applies to shorter tests held during the term.

Purposes of Examinations

Exams serve three main educational purposes:

- To test your factual knowledge;
- To test your ability to **synthesize** material and apply the concepts, methods, and problem-solving skills learned throughout a teaching session;
- To test your ability to explain and justify your informed opinion on some specific topic.

These reasons for exams give some indication of the sorts of things an examiner is likely to be looking for when marking a test. Some tests may serve only one of these purposes. For example, short-answer and multiple-choice tests may examine only your ability to recall information, whereas an exam consisting of essay questions or an oral examination of a thesis, will serve all three purposes.

Types of Examinations

Examinations in geography and the environmental sciences usually fall into one of four categories (or combinations thereof) (see Table 12.1).

Of these forms of exams, the closed-book is most common. Consequently, the following discussion will focus on that model. Nevertheless, much of the general advice will apply to the other kinds of exams. Some specific guidance on other exam types is provided at the end of the chapter.

An important component of success in exams is good "exam technique." Technique can be broken into two parts: (1) preparation and (2) writing the exam.

Table 12.1 ◉ Types of examinations and their characteristics

	Closed-Book
Characteristics	Requires that you answer questions on the strength of your knowledge and ability to recall information. Consulting any material in the exam room other than that provided by the examiner for the purposes of the test is not permitted.
Sub-types	• Multiple-choice • Short answer • Essay-type
	Open-Book
Characteristics	You may consult reference materials, such as lecture notes, textbooks, and journals during the exam. Sometimes the sources you are allowed to consult will be limited by your examiner.
Sub-types	• Exam-room • Take-home
	Oral
Characteristics	Used most commonly as a supplement to written exams, or to explore issues emerging from an honours, master's, or Ph.D. thesis. You may have to give a brief presentation before participating in a discussion, during which you will have to demonstrate to the examiners an understanding of your written work.
	Online
Characteristics	Requires that you complete and submit your exam via a computer. Used increasingly in distance-education courses. May consist of any of the exam types described above—with the exception of the oral exam.

How to Prepare for an Examination

Exams can be stressful, but anxiety can be managed and reduced with good study skills and preparation (Tracy, 2002; Van Blerkom, 2009). Well-prepared students are in a better position to achieve success on an exam than students who try to do all their studying just before the exam.

Ongoing preparatory activities during the term

Take notes

Success on exams is often based upon a foundation that is built throughout a course. Taking good notes during classes and from assigned reading is an important skill to develop. Exam review should consist of reading your notes and thinking about what you have learned. Notes provide essential links to and reminders of the material that was learned in classes or taken from readings weeks or months before an exam.

The use of PowerPoint by an instructor who makes the presentation available to his or her students may reduce the amount you need to write down during a class. However, having such material should not be viewed as an invitation to avoid taking *any* notes. Taking notes ensures that you stay engaged. Use the presentation notes as a starting point and take the opportunity to add material or make clarifications. These additions will enrich the base material and will help you later to recall what actually was discussed.

Here are some ideas for taking lecture notes without the benefit of having the professor's prepared notes in advance:

- *Write down major points (headings, main arguments, key facts) that will trigger your memory of the material when you are preparing for an exam*. If you realize that several minutes have passed and you have written nothing down, make a point of summarizing what took place or what was discussed in that period.
- *Supplement major points with a selection of more detailed information, examples, and anecdotes that will remind you of the broader themes of the class.* Students often make the mistake of thinking they have to write down *every* word that is spoken. But this takes away from actually hearing and understanding what is said. Not everything that is spoken has equal importance. Instead of trying to write down everything, write less but absorb more.
- *Write clearly and efficiently*. Notes have little or no use later if you are unable to decipher them. Again, writing less can allow you to write more

slowly and more clearly. Notes must make sense later but do not have to be highest-quality writing. Develop a set of common abbreviations, and learn which *non-essential* words you can usually omit without losing the meaning. These methods allow you to write more efficiently and to listen more.

When taking notes from readings, you should focus on synthesizing important passages, making connections that improve your understanding, and recognizing important statements, facts, or terms. Some people prefer to use a highlighter instead of writing out notes. This can be useful for identifying important material for later review, but it is not effective for synthesizing and making connections. For those tasks, you need to be more actively involved in processing and making sense of the material. Moreover, if most of the text has been highlighted, how will you be able to tell which parts are more important to review later? Just as in class, not everything in a textbook or article has equal importance, so if you prefer to highlight, learn to be selective.

Taking good notes is essential for exam review later.

Review course material regularly

This is one of the most difficult things to do in preparing for an exam because it requires commitment and follow-through. It is also very important. Review course material regularly as the term progresses, e.g., after each class, beginning on the very first day of classes. You might review by rewriting or rereading your lecture notes or by keeping up-to-date with note taking from assigned readings. Not only will this help you remember material when the time comes to write the exam, but it will also make it easier to understand lecture material as it is presented throughout the term. That can be a major benefit when it is time to complete the exam. Chapters 1 and 2 of Fred Orr's book entitled *How to Pass Exams* (2005) offer detailed guidance for organizing revision during the term. You might also find it helpful to keep in mind Dixon's (2004, pp. 152–153) sage counsel:

> [If] you have worked out what the key questions are in each course you are taking; have looked at old exam papers [if available] and paid attention to the course documentation; have attended and paid attention to your lectures and seminars; have made useful and well-organized notes; and have read around the subject and thought carefully about what your line is on the central issues; then you do not have too much to worry about.

Activities for successful exam preparation during review period

Find out details about the examination

Many professors will inform their class about the nature of the exam sufficiently far ahead to permit students to tailor their preparations accordingly. If you have not been given this information, ask your professor what kinds of question you may be asked on the exam, how much time will be allowed, what materials you will need, and so forth (see Box 12.1). Preparing for a multiple-choice exam is different from preparing for an essay-type exam, where there is greater scope for presenting your ideas or opinions, so it is important to know in advance which type of exam your professor will be using.

> *Reduce anxiety by finding out as much as possible about the exam.*

Seek some direction about how you might focus your preparation. Listen for clues from your professor about exam content. Professors often give thinly disguised hints about the content of an exam throughout the term, and sometimes the hints are not so thinly disguised!

Friedman and Steinberg (1989, p. 175) suggest that you should try to anticipate which questions and topics might be covered in the exam. Although this can be helpful, it can also be a risky game if you neglect to review some sections of the course material. Unless you have been told that the exam will concentrate on

Box 12.1 • Questions to ask about your examination

- How much does the exam contribute to the final grade?
- How long will we have for the exam?
- How many questions are on the exam?
- Is there a choice of questions to answer within sections (e.g., either/or options or a choice of a certain number of questions from a list)? Are any questions compulsory?
- What kind of questions are they, that is, multiple-choice, short answer, or essay-type?
- How are the marks allocated among the questions?
- Is there any information provided in addition to the exam, such as mathematical tables?
- Will material not covered in lectures and tutorials be examined?
- Can we use a calculator or dictionary?
- Are we allowed to take in any notes or books?
- Where is the exam and when is it being held?
- What is the procedure if I am sick on the day of the exam?

Source: Adapted from Hamilton (1999, p. 19).

specific topics, it is usually better to have a good overview of all the material covered in a course. Broad comprehension means that you should be able to tackle competently any question you encounter. If you narrow the scope of your revisions, you are gambling with your grade.

Reviewing for a limited range of topics covered in a course is a risky strategy.

Find a suitable study space

Arrange a comfortable, quiet, well-lit, and well-ventilated place where you can study undisturbed. If possible, make sure it is a place where you can lay out papers and books without risk of their being moved or lost. Some people find the library to be a good place to study, while others find that home is best. Whichever study space you choose, make sure you minimize distractions. At home, disruptions might include television, the Internet, and family members or roommates. At the library, being interrupted by friends is possible.

Set and maintain a study schedule

Once you have found out the dates and times of your exams, prepare a study schedule. This schedule should allocate specific days or parts of days to the review of specific topics in each course. When you make up your study schedule, think about those times when you work best and schedule your study periods accordingly. Although other activities and responsibilities are important, even during exam periods, set up the timetable so you can be doing those sorts of thing when you tend to feel mentally flat or slow. Kneale (1997) gives tips for time management during the term, which can be adapted easily to setting priorities during exams and developing an exam study schedule.

Make a study timetable and keep to it.

If your exams are weighted differently from one another or if you are performing better in one course than another, you should devote extra time to specific courses as appropriate. You may have become accustomed to a weekly pattern during the regular term and may find the absence of that structure during the exam period disconcerting.

Having a study schedule, and sticking to it, will help you make effective use of your time and avoid poor exam performance. You don't want to look back with regret on wasting time you had available for exam preparation.

Concentrate on understanding, not memorizing

On most written exams, you will be required to demonstrate an understanding of the subject material rather than to regurgitate information. Consequently, your preparation should focus on comprehension first and facts second. Be sure you understand what the course was about and the relationships between content and overall objectives.

When you have a grasp of the course objectives, you will be in a better position to make sense of content. Having a conceptual framework upon which you can hang substantive material will also allow you to answer the exam questions in a critical and informed manner. To develop this understanding, check the course outline or syllabus, lecture notes, practical exercises, and textbooks. Try to draw a concept map that indicates the key ideas and relationships between them. You may also find it helpful to condense your notes from, say, 30 pages to 6 and then to 1 page. Then expand them again without referring to the original notes. This process of condensing and elaborating on material should help you to develop a good understanding of the subject matter.

Usually it is better to study for meaning than for detail.

Vary your review practices

To add depth and to consolidate your understanding of course material, use various means of studying for a topic (Barass, 1984). Pose questions to yourself, solve problems, organize material, make notes, prepare simple summary diagrams, and read your notes thoughtfully.

Practice answering exam questions

Have a go at answering past exam questions or self-made questions under exam conditions. Working within time constraints is something that you should be ready for when you write the exam. Be sure you understand some of the key phrases and instructions that might appear, such as "discuss critically," "evaluate," and "compare and contrast." The Glossary at the end of this book may help to clarify some of these terms.

One good reason for practising exams under exam conditions has to do with changing technology. You probably do most of your in-term writing outside class on a computer. In an exam you may be asked to write by hand for up to three hours. Can you do that? Do your fingers and hands become stiff or tired? Computers allow you to move paragraphs around and make revisions quickly. Paper and pen do not offer that freedom, and you have to plan your writing much more carefully. You may also have come to rely on spelling and grammar tools in your software that are not available when you write by hand.

Seek help if you need it

If there is a topic that you do not understand, ask your professor for help. If you are having emotional, medical, or other difficulties that affect your studying, speak to the professor or to a counsellor. Most professors and institutions are willing to acknowledge genuine problems that may affect a student's ability to perform during an exam. If you feel that your condition will truly result in an unrepresentative result on an exam, be sure to consult your professor in advance. Seeking compensation for a poor performance *after* the exam is rarely successful.

Maintain your regular diet, sleep, and exercise patterns

Do not make the mistake of popping caffeine tablets and staying up into the early hours of the morning cramming information into your overtired brain. Radical changes to your lifestyle are likely to increase levels of stress and may adversely affect your exam performance. If you are in the habit of exercising regularly, keep doing that. Most people find that exercise perks them up, makes learning easier, and enhances exam performance.

> ***Good exam preparation includes looking after yourself.***

Preparations on the day of the examination

Pack your bag

Make sure you have your student identification card (in most universities you are required to present your ID in order to enter the exam) and any materials that are required or permitted, such as pencils, pens, ruler, paper, eraser, watch, and calculator. If it is an open-book exam, remember to pack the right books and notes. If the weather is hot, you might also want to pack a drink; or, if the room will be cold, remember to wear warm clothing or take an extra sweater.

Get to the right exam in the right place at the right time

Be sure that you know the scheduled time of the exam. You should also confirm the exam's location. When you are planning your departure for the exam, allow for the possibility of traffic delays, late buses, and bad weather. If you do miss the exam for some reason, contact your professor immediately.

Use the washroom

Before you enter the exam room, go to the washroom! It is astonishing how often students request a washroom break within the first half-hour of an exam. Asking to make more than one washroom visit during an exam may be viewed as suspicious and unnecessary by the **invigilator**. It would be neither unreasonable nor irresponsible for an invigilator to consider that the student may be leaving the room for reasons other than just using the facilities.

Techniques for Success in Examinations

Before an exam almost everyone feels tense and keyed up. However, if you have studied properly and know what kind of exam you will be writing, the anxiety you feel will probably help you perform at a higher level than if you were quite blasé about the whole affair. Breathe deeply and stride into the exam room with a sense of purpose. You know your stuff, and you know what the course was about. Here is the opportunity to prove it!

Steps to take before answering your first question

Make sure you have the entire exam

Be sure you have all of the exam's pages, questions, and answer sheets. On rare occasions, printing or collating errors may occur. If you believe this may have happened, check with the invigilator. Look on the back of every page of the exam to see if there are more questions hiding there.

Read the instructions carefully before beginning

Confirm the length of time of the exam, which questions must be answered, in what sections you have a choice of questions to answer, and how many marks each question is worth. Create a time allocation schedule (described below). Such an overview period will allow you to settle down and get a sense of what is required of you in the overall exam.

It is wise to *repeat* this overview process quickly after answering the first question (or several questions in the case of a short-answer exam) to confirm that your time allocation schedule is accurate. Professors find it most disheartening to mark an exam paper by a capable student who has not followed instructions (e.g., has not answered the right number or combination of questions). It is even more upsetting to be that student.

Work out a schedule

Calculate the amount of time you should devote to each question. This can be calculated approximately on the basis of marks per section or question. Table 12.2 gives an example of a schedule for an exam with sections of different length. Each section in this example may include one or more questions.

Table 12.2 ◉ Suggested time allocation for a three-hour examination

Task	Proportion of Exam Total Value	Time Allocation
Exam overview	n/a	10 minutes
Section 1	5%	8 minutes
Section 2	10%	16 minutes
Section 3	15%	24 minutes
Section 4	20%	32 minutes
Section 5	50%	80 minutes
Exam review	n/a	10 minutes
Total	**100%**	**180 minutes**

There is no time in this schedule allocated to breaks. Every so often you can stop, stretch, refresh your mind, and collect your thoughts—but this would have to be done within a section's time allocation. Reviewing or proofreading your answers at the end can also be a valuable use of time.

It is most important that you not only allocate your time carefully but *adhere to your timetable*. It would be poor practice, using the example in Table 12.2, to spend 30 minutes on Section 1 (that is, to spend 17 per cent of your time for 5 per cent of the exam value). This applies even if (to your credit) the answers are extremely well-written, complete, and in-depth. The point is obvious but worth stating: extra time spent on any question deprives you of time on others. Disciplined time management may be difficult, but it is undeniably a key to exam success.

Plan your exam.

On the other hand, you might sometimes improve your performance by using the full allocation of time on each question in order to write better answers rather than rush through to the end of the exam. Unused time could be used to rewrite or review answers at the end. There is a tendency for students to leave the exam when they have reached the end, regardless of how much time is left. To avoid this ineffective use of time and the feeling that you should leave because you have reached the end, stick to your schedule.

Advice for questions requiring written answers

Read each question carefully

During the exam overview period, read through the questions and make notes. Take note of significant words and phrases and underline them. Jot down ideas that spring to mind as you look over the questions. Use the time for a brainstorming session. After several hours of answering exam questions, you may have completely forgotten the brilliant ideas you had for questions toward the end. Your notes will trigger your memory.

Plan your answers

Do not make the mistake of rushing into your answers. Essay questions in an exam need to be approached in much the same way as essay assignments (see Chapter 3). Work out a strategy for your answer to each question. Use the ideas you jotted down during the overview as the basis of an essay plan for each answer. Make a rough plan of each answer before you begin writing. This might all be done on separate pages in your examination answer booklet. Take care to distinguish essay plans from final answers in your answer booklet by putting a pen stroke through the plan when you are finished.

Begin with the answers you know best

Number your answers. There is usually no requirement for you to answer questions in any particular order. It is often helpful to tackle the questions you consider to be easiest first to build up confidence and momentum. Furthermore, if you have the misfortune to run out of time, you will at least have submitted your best work.

Answer the question that is asked

Surprisingly, the most common mistake people make in exams is not answering the question that was asked, but instead spewing forth a prepared answer on a loosely related topic (Barass, 1984, p. 156; Friedman and Steinberg, 1989, p. 175). Your professor wants to know what you think and what you have learned about a *specified* topic. A right answer to the wrong question will not get you very far.

Examiners prefer focused, concise, and careful answers—not long-winded or wordy answers.

The examiners are assessing your understanding of particular topics. Do not try to trick them; do not try using the "shotgun technique," by which you tell all that you know about the topic irrespective of its relevance to the question; do not try to write lots of pages in the hope that you might fool someone into believing that you know more than you do. Concentrate instead on producing focused, well-structured answers. *That* will impress an examiner.

Attempt all required questions

It is foolish to leave any questions unanswered. In the worst case, make an informed guess. If you find that you are running out of time, write an introduction, outline your argument in note form, and write a conclusion. This will give the marker some sense of your depth of understanding and you may be rewarded appropriately.

Grab the marker's attention

People marking exam papers usually have many students to assess. They do not want to see the exam question repeated verbatim as the introduction to an essay answer. Nor do they want to read long, rambling introductions. Instead, they will want you to capture their attention with clear, concise, and coherent answers. Spare the padding. Get to the point.

Emphasize important points

You can emphasize key points by underlining words and by using phrases that give those points emphasis, such as "The most important matter is . . .," A leading cause of . . . ." You might also find it useful to use headings in essay answers to draw attention to your progression through an argument. Your examiner will certainly find headings useful. If used judiciously, bulleted or numbered lists and labelled diagrams can also be helpful.

Support generalizations

Use examples and other forms of evidence to support the general claims you make (Dixon, 2004; Friedman and Steinberg, 1989). Your answers will be more compelling and will signal your understanding of the course material more effectively if they are supported by suitable examples drawn from lectures and reading (remember the importance of taking notes?) than if they rest solely on bald generalizations.

Write legibly and comprehensibly

Examiners dislike having to decipher writing. They will find it difficult to follow someone's argument if they have to keep stopping to decipher hieroglyphics masquerading as the conventional symbols of written English. Try to write legibly. If you have problems writing in a form that can be easily read, write on alternate lines or print to ensure the examiner is able to interpret

Practise writing legibly by hand before a long exam, especially if you are more accustomed to typing.

your work. Keep your English expression clear, too. Errors of punctuation and grammar may divert the examiner's attention from the positive qualities of your work. Many problems of expression can be overcome by using short sentences. These also tend to have greater impact than long sentences.

Leave space for additions

Leave space for adding material if time allows. This is particularly important if you have been prudent enough to leave yourself some time for proofreading. Often, people find that they recall material for one question while they are answering another. It is useful to have the time and space to add those insights.

Keep calm

If you find that you are beginning to panic or that your mind has gone blank, stop writing, breathe deeply, and relax for a minute or two. A few moments spent this way should help to put you back on track. Do not give up in frustration and storm out of the exam room. Why run the risk of working out a way through a problem after you given up your chance to complete the exam?

> *Don't leave an examination early in despair. Stay and think. You won't be allowed back in once you leave.*

Whatever you do, do not cheat! Exams are carefully supervised. Copying and other forms of academic dishonesty in examinations do not go unnoticed. Your entire exam mark and course grade are at stake if you are caught cheating at any point during the exam.

Proofread your answers

Allow yourself time to proofread your answers. Check for grammatical errors, spelling mistakes, and unnecessary jargon. You may also find time and opportunity to add important material you missed in your first attempt at the question.

Advice for multiple-choice examinations

Aside from familiarity with the test material, good results in multiple-choice exams depend, in part, on being conversant with some of the peculiarities of this form of exam. As the following points suggest, there is much more to success in a multiple-choice exam than simply choosing a letter from a list of A to E.

For each question, be sure to follow these tips:

- Read all the answer options before selecting one.
- Cut down your list of choices by eliminating answers that are clearly incorrect.
- Avoid answers that incorporate unfamiliar terms (Burdess, 1998, p. 57; Northey et al., 2012, p. 204). Examiners will sometimes use technical words as sirens, luring you in. Your preparations should have made you familiar with the relevant terms that could form part of correct answers.
- If you are not confident about selecting an answer from among the options provided, choose the option you judge closest to being correct.
- If you are truly stuck and face a choice between two or more answers, take a guess. Unless you have been advised that penalties are imposed for incorrect answers, answer every question. For a question with five options, you have a 20 per cent chance of success. If you cannot favour one answer over another, make the most educated guess possible.

For the exam as a whole, follow these suggestions:

- Go through the exam answering all those questions you can answer easily. If there is a question you find difficult, move on and return to it later if you have time.
- Answer each question independently without being intimidated or led astray by an emerging pattern of answers. For example, it does not make sense not to select the answer you think is right to avoid a certain pattern on the answer sheet. If you have decided that one answer is your best choice for that question, select it.
- Do not "over-analyze" a question. If you have completed a multiple-choice exam and are reviewing your answers, be cautious about changing your original answer unless you are certain it is wrong. Have faith in your initial response to the question, particularly if you are in doubt.

Advice for oral examinations

Formal oral exams are relatively rare in geography and the environmental sciences, and hence most people do not get the opportunity to practise them as they might a written or multiple-choice exam. Except in small classes, a written exam is far more efficient for professors than an oral exam. Honours and graduate students often have oral exams at the end of their research projects. Because of the unfamiliarity, an oral exam can be quite intimidating. But if you think about it, the oral examination is simply a formalized extension of the sort of discussion you may have had with colleagues, friends, and supervisors about the subject you are studying. As such, it should not be too daunting a prospect.

To prepare for an oral exam, think about its probable aims. If you do not know what those aims are, ask your professor or supervisor. Generally, the examiners will want you to fill in more detail or provide clarification about material you wrote in a paper or thesis. They may want to test the depth of your understanding of certain topics. The examiners might also wish to use an oral exam as a teaching and learning forum. They may want to encourage you to think about alternative ways in which you might have approached your topic and to discuss those alternatives with them. If you can, try to look forward to an oral exam as a potentially rewarding opportunity to explore a subject about which *you* may be the best informed.

As specific advice for completing an oral examination, some of the following points may be useful.

- If the exam is about a thesis or some other work you have written, read through that work, taking a few notes on its key points and otherwise refreshing your memory about it.
- As the date of your exam approaches, conduct a search of recent journals to see if any relevant papers have been published. Your examiners may not yet have read those papers and this will give you an opportunity to impress them (Beynon, 1993, p. 94).
- Enlist the help of a friend or family member. Try to think of the sorts of things the examiner might ask you to demonstrate, and practise some brief, clear answers. Box 12.2 lists some questions commonly asked in oral exams. In trying to make your answers to these questions clear to your friend or relative, you may discover areas that need further revision.
- At the exam itself, present yourself in a way that is both comfortable and suitable to the formality of the occasion.
- If you have been asked to make an introductory presentation, keep it short and to the point. Usually no more than 10 per cent to 20 per cent of an oral exam period will be allocated to an opening presentation. Emphasize the main points, from your paper or thesis, such as key research objectives and findings. Avoid getting bogged down in detail.
- During the question period try to display relaxed confidence. To do this, maintain eye contact with your examiners, sit comfortably in an alert position, and do not fidget. Remember to speak clearly. Try to formulate brief but clear answers.

Box 12.2 ◉ Common questions in oral exams

A number of questions appear regularly in oral exams. Give them some careful thought before the exam.

- Why did you select the particular research question?
- How does your work connect with previous studies?
- Does it say anything different? Does it confirm previous work?
- How does your work fit into the field of geography or environmental science?
- Why did you select your methodology? Are there any weaknesses in the approach you adopted?
- What are some of the sources of error in your data?
- What practical problems did you encounter? How did you overcome them?
- Are any of the findings unexpected?
- What avenues for future research does your work suggest?
- What are the particular strengths or weaknesses of your research?
- What have you learned from your research experience? What would you do differently if you had to repeat the study?

Before an oral exam, prepare your answers to questions you are most likely to be asked.

- If you do not understand a question, say so. Ask to have the question rephrased. It may not be you who has the problem; it is quite possible that the question as phrased does not make sense.
- For longer questions, write down key terms while the question is being asked. Some questioners will announce that they are going to ask a two-part question. Making a note about each part will enable you to address both parts clearly without having the question repeated.
- Take time to think about your answers. Do not feel obliged to rush a response.
- In most exams, as in a discussion, you should feel free to challenge the examiners' arguments and logic with your own evidence or ideas, but be prepared to give consideration to their views.
- If it appears to you that the examiners have misinterpreted something you have said or written, let them know precisely what you did mean.

Advice for open-book examinations

The "open-book" exam is deceptive. At first, you might think, "What could be easier than a test where I am allowed to open my book to look for the answers?" Later—perhaps as late as during the exam itself—you may come to realize that these exams can be a trap for inexperienced students. Open-book exam questions rarely ask for factual answers that you can simply look up. The key to success is to know your subject very well. The open-book exam simply allows you to obtain specific examples, references, and other material that might support or enrich *your* answers to questions. You still have to interpret the question and devise an answer. *You* must produce the intellectual skeleton upon which your answer is constructed, and then place upon that the flesh of personal knowledge or examples and arguments that may be drawn from reference material. If you do not understand the background material from which the questions are drawn, you may not be able to do as well as you should.

The following may be useful additional advice in preparing for an open-book exam:

- Study the material generally as you would for a closed-book exam.
- Prepare easily understood and accessible notes for ready reference.
- Become familiar with the reference materials you are planning to use. If appropriate, mark sections of texts in a way that will allow you to identify them easily, e.g., with a highlighter or sticky notes, because you do not want to waste valuable exam time searching for supporting information. Do not mark library books or ones belonging to other people!

Advice for take-home examinations

While a take-home exam may appear to offer you the luxury of time, preparation remains a key to success. Do not make the mistake of squandering your time trying to get organized *after* you have been given the exam question(s). Not only will you be wasting valuable time but the preparation is likely to be hurried and inadequate. Here are some suggestions to help you prepare for a take-home exam.

Like an open-book exam, a take-home exam allows you to write better answers by giving you access to reference material. Be sure to have at your disposal suitable reference material that you are familiar with. In short, you should have read through, taken notes from, and highlighted a sufficiently large range of reference materials to allow you to complete the exam satisfactorily. Arrange all the reference material in a way that allows you to find specific items quickly.

Take-home and open-book exams require careful preparation.

Set up a timetable by allocating periods of time for each question in the exam. Even if you are given 48 hours for the exam, you obviously will not be working on the questions for all that time. Your aim should be to produce concise, carefully considered, well-argued answers, supported with examples where appropriate. You are not meant to be writing for the entire time.. Instead, think carefully and focus clearly on the questions asked. You also need to consider if you will have other academic and extra-curricular responsibilities during the period in which the take-home exam has been assigned.

Set limits on the time and resources you will use in a take-home exam.

Advice for online examinations

Much of the advice about preparing for and writing exams that has been mentioned earlier in this chapter also applies to this form of exam. Make sure you know what resources you will be allowed to use, or restricted from using, during the exam.

If you are completing your exam in a room with other students, they may be working on questions in a different order than you. Some testing programs set this up to discourage cheating. Be aware of your own progress and schedule in relation to the time available. Do not be intimidated or discouraged by the apparent progress of others.

When you finish your exam, be sure you have saved a copy of your answers (if possible) and have received an acknowledgement of your submission. This might take the form of an on-screen message or confirmation number.

Further Reading

Barass, R. (1984). *Study! A guide to effective study, revision and examination techniques.* London: Chapman and Hall.

- Chapters 12 and 13 of this book comprehensively describe the steps in preparing for and taking various types of examination. This is a very helpful reference book.

Burdess, N. (1998). *The handbook of student skills for the social sciences and humanities* (2nd ed.). New York: Prentice Hall.

- This book includes two dozen useful pages on preparing for and writing exams.

Clanchy, J., and Ballard, B. (1997). *Essay writing for students: A practical guide* (3rd ed.). Melbourne: Longman Cheshire.

- The chapter on exam essays includes a short discussion on ways of preparing for examinations.

Friedman, S.F., and Steinberg, S. (1989). *Writing and thinking in the social sciences.* Englewood Cliffs, NJ: Prentice-Hall.

- Chapter 13 discusses, in some detail, 10 handy hints to help improve performance in written exams.

Hay, I., and Bull, J. (2002). Passing online exams. *Journal of Geography in Higher Education, 26*(2), 239–244.

- This article offers an extended discussion of the material presented in this chapter on ways of succeeding in online exams.

Northey, M., Knight, D.B., and Draper, D. (2012). *Making sense: Geography and environmental sciences* (5th ed.). Toronto: Oxford University Press.

- Chapter 12 has a short section on preparing for different types of examination.

Orr, F. (2005). *How to pass exams.* Sydney: George Allen and Unwin.

- This is a comprehensive book on exam preparation and performance. Most emphasis is given to intellectual, physical, and psychological means of preparing for examinations.

Tracy, E. (2002). *The student's guide to exam success.* Buckingham, UK: Open University Press.

- This book presents techniques for preparing for exams and reducing anxiety to improve exam performance.

Van Blerkom, D.L. (2009). *College study skills: Becoming a strategic learner* (6th ed.). Boston: Wadsworth Cengage Learning.

- As a general guide for new post-secondary students, this book includes four chapters about tests and exams, with separate chapters on objective exams and essay exams.

Appendix: Sample Paper for a Short Written Assignment

This appendix contains two versions of a short paper that might be submitted for a written assignment. The first paper is the "poor version." It has been written intentionally to be flawed and to contain errors. This version might be considered acceptable as a first draft that could be transformed, through careful attention and revision, into a paper suitable to be submitted for grading. The paper has been prepared to highlight some issues with clarity, structure, citations and referencing, spelling, and proofreading.

The second paper is the "better version," written on the same topic so that you can compare the two versions directly. When you read the better version after reading the poor version, can you see the differences even though both writers were discussing the same topic? Hopefully you will agree that the better version is a much stronger paper and it will inspire you to work on developing your writing skills.

Numbered discussion notes in the right-hand column will help you understand the flaws and errors in the poor version. In the better version the discussion notes highlight some of its strengths, particularly when viewed in comparison with the poor version.

Bear in mind that not all of the matters discussed in this book can be addressed in this example. Only a limited selection of positive and negative examples of writing for an assignment are shown and discussed. Given the short length of the paper, the use of headings was not deemed necessary. The author-date (in-text) citation style is used instead of footnotes or endnotes.

Sample Paper: Poor Version

Ice-Jam Floods (1)

(2) A flood is when a river spills over on to the floodplain. A flood happens when the volume of water supplied from upstream exceeds the capacity of the river (3). Floods are caused by blockages in the river channel which create a dam and the normal flow rate is impeded and the inflowing water is backed up. Ice jams are a common cause of blockage (Beltaos) (4). Constriction points are common sites for ice jams because the

1 The title is a key term but is not specific enough to indicate what the paper is about. The paper is a significantly more focused explanation of one aspect of ice-jam floods.

2 The first paragraph is an unfocused mix of information that does not clearly define the paper's topic. It begins by discussing floods in general, and ice jams specifically are not mentioned until the fourth sentence. Extraneous material for an introduction is included with the facts about the Mackenzie River. These issues are indicative of insufficient attention to structure and a lack of careful revision.

3 A citation for the definition of a flood is recommended here. Although people generally know what a flood is, this is a specific definition that is not common knowledge.

4 Citation should be "(Beltaos, 2000)."

blocks of ice become wedged against the banks and each other more easily (Saade, 1995) 5. Ice jam floods 6 occur on higher latitude rivers in the northern hemisphere. Environment Canada's website 7 says that the MacKenzie river 8 has annual ice jam floods. The Mackenzie River 9 is in Yukon Territory and empties into the Bering Sea 10. Ice jam floods occur on northward flowing rivers 11. This paper is about ice-jam floods and why their 12 common in the Northern Hemisphere 13.

Gerard said that a particular sequence of natural events normally coincides with an ice-jam flood (1990) 14. First, 15 ice cover must form on the river. Two, discharge rises in the headwater regions of the drainage basin. This occurs with snowmelt in spring or during a midwinter thaw period. Third, the meltwater converges in main-stem channels while the river downstream is still frozen. Geographic characteristics of a drainage basin and stream network determine the occurrence of an ice-jam flood.

In a summary of Russian physical geography, Armstrong Rogers & Rowley (1978) 16 stated that river drainage is almost all northwards so the fact that the headwaters of most is in the south leads to extensive annual flooding as the spring melt progresses northward 17. Melting of snow cover is more likely to happen first at lower latitudes (Rouse et al., 1997) before river ice breaks up at higher latitudes. Meltwater converging from the southern parts of a drainage basin is blocked by the ice-jam further north. (O'Connor et al., 2002; Bone, 2003) 18

5 *Saade* is misspelled; it should be *Saadé*. Also, the source has more than two authors [see reference list] so citation should be "(Saadé et al., 1995)."

6 Correct terminology is "ice jam" with no hyphen, but "ice-jam flood" with a hyphen. "Ice jam flood" is incorrect.

7 Environment Canada's website is mentioned but no citation is given. An entry is included in the List of References but there needs to be a corresponding in-text citation to make the connection.

8 "MacKenzie river" should be "Mackenzie River."

9 Inconsistency in appearance of a name indicates a lack of careful proofreading. In the previous sentence, the reader wrote "MacKenzie river" but here it is written as "Mackenzie River." Similarly, there is inconsistency throughout the paper in whether the N and H in Northern Hemisphere are in upper or lower case.

10 Facts are incorrect: the Mackenzie River is in the Northwest Territories, not Yukon Territory, and it flows into the Beaufort Sea, not the Bering Sea. A reader who recognizes these errors is more likely to question other facts in the paper.

11 As it is written, this carelessly worded sentence suggests to the reader that ice-jam floods occur only on northward-flowing rivers. The point made in the better version is that ice jams can occur on rivers flowing in any direction but northward-flowing rivers are more susceptible.

12 Spelling mistake: *their* instead of *they're*. However, contractions (*they're*) should be avoided in academic writing anyway.

13 At last—something that resembles a thesis statement. However, it does not need to be written as "This paper is about . . . ." Also, it is too vague. Compare with the thesis statement in the better version of this sample paper.

14 Author and year of source publication should appear adjacent, as in "Gerard (1990)."

15 A list of items is presented using inconsistent wording: "*First . . . Two . . . Third*." To be consistent it should be "*First . . . Second . . . Third*."

16 Citation should be Armstrong et al. (1978).

17 The statement from Armstrong et al. (1978) is actually a quotation, so it should be surrounded by quotation marks.

18 Citations are included in the sentence in which the corresponding material appears. Therefore place the citation before the period that ends the sentence, not after it.

In the higher-latitude regions of the Southern Hemisphere which have the required climatic conditions, there is a lack of land surfaces with southward-flowing rivers. [19]

Ice jam floods are more common on rivers in higher-latitude regions of the Northern hemisphere because climatic patterns cause snowmelt in the south resulting in northward flow of water while there is still ice in the river further north. [20] However, ice jams can occur anywhere [21].

List of References 

Beltaos, Spyros. 2000. Hydrological Processes, 14(9), 1613-1625. Advances in river ice hydrology.

R.G. Saadé, A.S. Ramamurthy, and M.S. Troitsky, (1995). Numerical Modelling of Ice Jam Resistance to Main Channel Flow. *International Journal for Numerical Methods in Fluids, 21*(11), 1109–1120.

Rouse, W. R., et al. (1997). Effects of climate change on the freshwaters of arctic and subarctic North America. *Hydrological Processes*, 873–902.

Bone, R.M. *The Geography of the Canadian North* (2nd ed.) Oxford University Press, Don Mills, ON, 2003.

Christopherson, Robert W. (2013). *Elemental Geosystems* (7th ed.). New York, Pearson.

Gerard, R. Hydrology of floating ice. *Northern Hydrology: Canadian Perspectives*. Saskatoon, SK: National Hydrology Research Institute (pp. 103–134).

Environment Canada. (2013). www.ec.gc.ca/eau-water/default.asp?lang=En&n=155A4B17-1 (1 August 2013).

Armstrong, T., Rogers, G., and Rowley, G. (1978). *The Circumpolar North*. Methuen & Co.

O'Connor, J.E., Grant, G.E., Costa, J.E. (2002). The geology and geography of floods. In P.K. House, R.H. Webb, V.R. Baker, and D.R. Levish, *Ancient Floods, Modern Hazards: Principles and Applications of Paleoflood Hydrology*. 359–385.

19 Avoid writing one-sentence paragraphs. Furthermore, this sentence is an incomplete thought. In the better version this sentence is preceded by another sentence; together they provide a complete thought about why ice-jam floods do not occur in the Southern Hemisphere.

20 A moderately clear summary sentence is written. One might believe that this defines the writer's objective, but it is difficult to tell what the objective actually is because the title, introduction, and thesis statement are poorly written.

21 Another example of material that is poorly written and misleading. Ice jams do not occur anywhere; certain climatic conditions are required. The writer meant to say that northward-flowing rivers are more susceptible to ice jams, but if the climatic conditions are suitable they can occur on other rivers, too.

22 See how many formatting errors you can identify in this list of references—numerous mistakes were made. Pay close attention to details! To find out what the correct formatting should be, see the list following the better version of the sample paper.

Sample Paper: Better Version

Flow Direction of Rivers: An Important Control on the Frequency of Ice-Jam Floods (1)

(2) Ice-jam floods are phenomena that occur on many higher-latitude rivers in the Northern Hemisphere. (3) Certain climatic conditions are required to cause an ice-jam flood. Although ice-jams can happen on rivers that flow in any direction, the frequency of this type of flooding is greater on rivers that flow northward. Examining the nature and causes of ice-jam floods reveals why northward-flowing, higher-latitude rivers in the Northern Hemisphere are particularly susceptible to these events (4).

A common definition of a flood (Christopherson, 2013, p. 362) (5) is when the level of water in a river rises above the threshold of its banks and spills over onto the adjacent floodplain. This happens when the volume of water supplied from upstream exceeds the capacity of the river channel. There are various reasons why the channel capacity may not be sufficient to accommodate the supplied discharge. One reason is a blockage in the channel which acts as a dam to impede the normal flow rate and backs up the inflowing water. A specific type of blockage is an ice jam, which can occur at freeze-up or breakup (Beltaos, 2000). Constriction points, where a river channel narrows, are a common site for breakup ice jams (Saadé et al., 1995) because the blocks of ice become wedged against the banks and each other more easily. With the damming

1 The title uses several key terms to provide a succinct indication of what the paper is about.

2 After reading the introductory paragraph, the reader will know what the paper sets out to achieve. Bold statements are made, but evidence to support those statements is *not* provided here. Save the evidence for the body of the paper.

3 The reader may have been provoked into asking questions such as "Are ice-jam floods a phenomenon only on rivers in the Northern Hemisphere, not in the Southern Hemisphere?" The wording of this first sentence could lead readers to think that ice jams occur only in the Northern Hemisphere. Therefore the writer must take care to address that question at some point in the paper.

4 The last sentence of the introductory paragraph is the thesis statement. In the rest of the paper the reader will be expecting to see (a) information about the nature and causes of ice-jam floods, and (b) evidence supporting the claim that northward-flowing, higher-latitude rivers are particularly susceptible to ice-jam floods.

5 Although the citation is to material that is paraphrased but not quoted, a page number is given because it refers to specific material in the source.

effect of the ice jam, the water level rises and flooding onto the floodplain follows. Canada's Mackenzie River is an example of a northern river where ice-jam flooding occurs annually (Environment Canada, 2010). Beltaos (2000) 6 also described downstream flooding caused by the surge of water as a large volume of ponded water is suddenly released when the ice jam eventually breaks up.

6 To create some variety in the way citations are incorporated into the writing, here the author of the source was used as the subject of the sentence. More often, the citation is simply inserted at the end of the sentence that draws on the work, as in the first reference to the Beltaos (2000) article earlier in this paragraph.

A particular sequence of natural events normally coincides with an ice-jam flood (Gerard, 1990) 7. First, ice cover must form on the river so there is a cold climate limitation on where ice jams and the related floods occur. Second, discharge rises in the headwater regions of the drainage basin. This typically occurs with snowmelt in spring or during a mid-winter thaw period. Third, the meltwater converges in main-stem channels while the river downstream is still frozen. Although the increased discharge can begin to break up continuous ice cover on a river, blocks of ice can still become stuck, which causes the damming effect described above. Geographic characteristics of a drainage basin and stream network therefore play an important role in determining the likelihood of occurrence of an ice-jam flood.

7 The citation to Gerard (1990) refers to a particular sequence of natural events. The paragraph goes on to describe the sequence using the flags *"First . . . Second . . . Third"* to connect with the phrase "sequence of natural events." With this structure, only one citation is required because it is implied that the information in the paragraph comes from the Gerard (1990) source.

Given the geographic influences on the sequence of events causing ice-jam floods, it becomes clear why northward-flowing rivers in the Northern Hemisphere have greater susceptibility to this phenomenon. In a summary of Russian physical geography, Armstrong et al. (1978, p. 23) 8 stated that ". . . river drainage is almost all northwards . . . . so the fact that the

8 A direct quotation is used instead of the more frequent paraphrasing technique, in line with the advice to use quotations sparingly and for effect. In this quotation some of the source text is omitted and replaced with ellipsis points (. . .).

headwaters of most [9] [rivers] is in the south leads to extensive annual flooding as the spring melt progresses northward." Melting of snow cover is more likely to happen first at lower latitudes (Rouse et al., 1997) before river ice breaks up at higher latitudes. Meltwater converging from the southern parts of a drainage basin is blocked by the ice jam further north (O'Connor et al., 2002; Bone, 2003) [10] . Theoretically, a similar situation would occur in the Southern Hemisphere, but the required combination of factors do not exist for ice-jam floods to occur. In the higher-latitude regions of the Southern Hemisphere, which have the required climatic conditions, there is a lack of land surfaces with southward-flowing rivers. [11]

Floods caused by ice jams can occur anywhere that water is supplied from the drainage basin before river ice cover downstream has melted [12]. However, the conditions that cause ice-jam floods are more prevalent on rivers in higher-latitude regions of the Northern Hemisphere. This is because climatic patterns cause snowmelt in the south, resulting in northward flow of water while there is still ice in the river further north [13].

9 Because some text is omitted, the word "rivers" is inserted in square brackets so that the sentence still makes sense. Without it, the reader would be left asking "The headwaters of most *what* is in the south?"

10 Two sources are cited for the same piece of information. Citing independent sources for the same fact can be used to highlight that the information is corroborated in multiple sources, instead of just appearing in isolation in one source. Writers often do this to emphasize the presence of broader themes or major issues that are discussed in a body of literature.

11 This paper makes the claim that rivers flowing northward into cold regions are *more susceptible* to ice-jam flooding. It does *not* claim that ice-jam flooding occurs only on rivers flowing northward into cold regions. Similarly, the title of the paper says that flow direction is *an* important control on frequency of ice-jam flooding and does not imply or state that it is the *only* control.

12 No citation is given for this statement. There may be a source in existence which does include this information, but the writer's own awareness of global geography is on show here. The statement qualifies as common knowledge not requiring a citation.

13 Note how the concluding paragraph relates back to the objectives and thesis statement presented in the introductory paragraph. However, the conclusion does not simply repeat the introduction. In its position at the *end* of the paper the conclusion briefly summarizes the evidence that was only alluded to in the introduction.

List of References

Armstrong, T., Rogers, G., and Rowley, G. (1978). *The Circumpolar North*. Methuen & Co.: London. [14]

Beltaos, S. (2000). Advances in river ice hydrology. *Hydrological Processes, 14*(9), 1613–1625. [15]

Bone, R.M. (2003). *The Geography of the Canadian North* (2nd ed.) Don Mills, ON: Oxford University Press. [16]

14 Complete book, by three authors.

15 Journal article, by single author.

16 Complete book, subsequent edition, by single author.

Christopherson, R.W. (2013). *Elemental Geosystems* (7th ed.). New York: Pearson.

Gerard, R. (1990). Hydrology of floating ice. In T.D. Prowse and C.S.L. Ommanney (eds.), *Northern Hydrology: Canadian Perspectives*. Saskatoon, SK: National Hydrology Research Institute (pp. 103–134). 17

Environment Canada. (2010). *Flooding Events in Canada—Northwest Territories*. www.ec.gc.ca/eau-water/default.asp?lang=En&n=155A4B17-1 (accessed 1 August 2013). 18

O'Connor, J.E., Grant, G.E., and Costa, J.E. (2002). The geology and geography of floods. In P.K. House, R.H. Webb, V.R. Baker, and D.R. Levish (eds.), *Ancient Floods, Modern Hazards: Principles and Applications of Paleoflood Hydrology*. Washington: American Geophysical Union (pp. 359–385). 19

Rouse, W.R., Douglas, M.S.V., Hecky, R.E., Hershey, A.E., Kling, G.W., Lesack, L., Marsh, P., McDonald, M., Nicholson, B.J., Roulet, N.T., and Smol, J.P. (1997). Effects of climate change on the freshwaters of arctic and subarctic North America. *Hydrological Processes*, *11*(8), 873–902. 20

Saadé, R.G., Ramamurthy, A.S., and Troitsky, M.S. (1995). Numerical modelling of ice jam resistance to main channel flow. *International Journal for Numerical Methods in Fluids*, *21*(11), 1109–1120. 21

17 Chapter, by single author, in edited volume with two editors.

18 Web page, with date of last modification (2010), accessed on 1 August 2013.

19 Chapter, by three authors, in edited volume with three editors.

20 Journal article, by 11 authors.

21 Journal article, by three authors.

Glossary

abstract Short summary of the objectives, methods, results, and central conclusions of a research report or paper. An abstract is usually about 100–250 words and appears at the beginning of the paper. It is designed to be read by people who may not have the time to read the whole report, who wish to get a quick impression of the content, or who are deciding whether the content is sufficiently interesting for them to read the entire document.

account Explanation of how something came about and why.

acknowledge Show recognition of or give credit to author(s) for previous work or ideas.

acknowledgements Section of a paper, article, or book in which the author recognizes and thanks the people and institutions to which he or she is indebted for guidance and assistance. May be incorporated into the preface.

acronym Word made up of the first letters of a group of words (e.g., SCUBA—Self-Contained Underwater Breathing Apparatus). Strictly speaking, a group of initials that is *not* pronounced as a word, such as CBC, is not an acronym but an *initialism*.

ad lib Improvise speech or performance.

analyze Explore component parts of some phenomenon in order to understand how the whole thing works. It can also mean "to examine closely." (Compare with *synthesize*.)

annotated bibliography A list, in alphabetical order by author's surname, of works (books, articles, etc.) on a specific topic. Each work is summarized and commented upon.

annotation Explanatory note or comment. May be used as a label for clarification on a graph or diagram, or to summarize and review a work briefly.

appendix Supplementary material accompanying the main body of a paper, book, or report. Is placed at the end of the document or sometimes at the end of a chapter. Consists of supporting evidence that would detract from the main line of argument in the text or would make the main text too long and poorly structured.

appraise Analyze and judge the worth or significance or something.

argue Present reasons to support or oppose an issue.

argument Debate that involves reasoning about all sides of an issue and offering support for one or more sides. In a debate, you will be asked to present a case for or against a proposition, giving reasons and evidence for your position. In an argument, you should also indicate opposing points of view and your reasons for rejecting them. An argument may be written or spoken.

arithmetic scale Scale on a graph in which the main intervals are separated by constant values. (Compare with *logarithmic scale*.)

assess Conduct an evaluation, investigating the pros and cons or validity of some issue or situation. You are usually expected, on the basis of your research and discussion, to reach some conclusion (e.g., whether some situation under consideration is right or wrong, fair or unfair).

author-date system System of referring to texts cited. Comprises two parts: in-text *citations* and an alphabetically ordered *reference list*. (Compare with *note system*.)

bar graph General name given to those graphs in which plotted values are shown in the form of one or more horizontal or vertical bars (column graph) whose length is proportional to the value(s) portrayed. (Compare with *histogram*.)

bibliographic details Information about a publication, usually consisting of the author's name, date of publication, title, publisher, place of publication, and, if relevant, the publication's volume number, issue number, and page numbers.

bibliography Complete list of works referred to or found useful in the preparation of a formal communication (e.g., essay, book review, poster, report). (See also *reference list* and *annotated bibliography*.)

blog Web application that allows one or more people to post news, information, musings, personal diaries, political commentary, photographs, etc., on a common web page. Many blogs have minimal editorial oversight and are used to express personal opinions; therefore, they should be used as an information source with extreme caution.

body Main part of a work that contains the detailed material foreshadowed in the introduction and summarized in the conclusion; presented to make a case or argument.

caption Explanatory text accompanying a figure or illustration.

cartogram Form of map in which the size of regions depicted is adjusted to represent the statistic being mapped. Although the physical sizes of places will be altered in the production of a cartogram, efforts are made to preserve both their shapes and their locations in relation to other places.

cartographic scale Statement of the relationship between distances on a map and corresponding distances on the ground. Cartographic scale may be displayed as a ratio, a graphic or bar, a representative fraction, or a verbal statement.

choropleth map Cross-hatched or shaded map used to display statistical distributions (e.g., rates, frequencies, ratios) on the basis of areal units, such as nations, states, and regions.

citation Formal, written acknowledgement that you have borrowed the work of another scholar. Whenever you quote the work of another person verbatim (i.e., reproduce word for word) or borrow an idea from another person, you must acknowledge the source of that information by using a recognized referencing system.

comment (on) Make critical observations about.

common knowledge Facts, observations, and information that may be mentioned by an author without requiring that a source be cited.

compare Discuss the similarities and differences between selected phenomena (e.g., ideas, places). (Often used in conjunction with *contrast*.)

concept Thought or idea that underpins an area of knowledge. For example, the concept of evolution underpins much of biology. The concept that new communications technologies "compress" distance is significant in geography.

conceptual framework Logic that underpins an argument or the way in which material is presented. A way of viewing the world and of arranging one's observations into a comprehensible whole. May be imagined as an intellectual skeleton upon which flesh, in the form of ideas and evidence, is supported.

concluding sentence Last sentence in a paragraph; lets the reader know the paragraph is complete.

conclusion Part of a talk, essay, poster, or report in which findings are drawn together and implications are revealed.

consider Reflect on; think about carefully.

continuous data Data that could have any conceivable value within an observed range. Thus includes fractional numbers, not just integers. (Compare with *discrete data*.)

contrast Give a detailed account of differences between selected phenomena. (Often used in conjunction with *compare*.)

corroboration Support or confirmation of an explanation or account through the use of complementary evidence. (Compare with *replication*.)

critical reading Reading of a text with an active approach and a skeptical attitude, i.e., examining and questioning the author's ideas instead of taking everything you read at face value.

criticize Express opinions about the strengths and weaknesses of something; back your case with a discussion of the evidence. Criticizing does not necessarily require that an idea be condemned.

critique See *criticize*.

data region That part of a graph within which data are portrayed (usually bounded by the graph's axes).

define Explain the basic points or principles of something so as to provide a precise meaning. Examples may enhance a definition.

demonstrate Illustrate and explain by use of examples.

describe Outline the characteristics of some phenomenon. Usually, the act of describing might be imagined as painting with words. There is no need to interpret or judge.

discourse Communication of thought by words.

discrete data Data that may be expressed in whole numbers only, such as animal and human populations. (Compare with *continuous data*.)

discriminatory language Language that treats people differently, or excludes them, on the basis of their sex, race, religion, culture, age, or disability where it is unnecessary or inappropriate to do so.

discuss Examine critically, using argument. Present one's point of view and that of others. Discussion may be written or spoken.

distinguish Make clear any differences (between two or more phenomena).

DOI Stands for Digital Object Identifier, a string of characters used to identify an electronic document or other object in the digital environment.

dot map Map in which spatial distributions are depicted by dots representing each unit of occurrence (e.g., one dot represents one person) or some multiple of those units (e.g., one dot represents 1000 people).

edit Revise and rewrite; prepare a work such as an essay for submission. Sometimes implies that some material is deleted.

ellipsis points Punctuation mark, written as three periods (. . .), placed in a quotation, for example, to indicate that words from the original source have been omitted.

endnote Short note placed at the end of a document or chapter and identified by a symbol or numeral in the body of the text. A textual "aside," an endnote provides a brief elaboration of some point made in the text whose inclusion there might disrupt the flow of text. In the *note system* of referencing, endnotes may also include details of reference material cited in the text. See also *footnote*.

essay Literary composition that states clearly what you think and have learned about a specific topic.

essay plan (or essay outline) Preparatory framework outlining the basic structure and argument of an essay.

et al. Abbreviation of the Latin phrase *et alii*, meaning "and others," used to shorten a list of authors' names in citations (e.g., Maxwell et al., 2003).

evaluate Appraise the worth of something. Say what the strengths and weaknesses are and which predominate. Make a judgment.

evidence Information used to support or refute an argument or statement. In forming an opinion or making an argument at university, you may need to abandon some practices that may have been considered satisfactory in the past. For example, it is not acceptable to say that "it is widely known that . . ." or "most people would say that . . ." since in these statements you have not provided any evidence about who the people are, why they say what they do, how they came to their conclusions, and so on. In other words, you need to present material that supports or refutes your claim.

examine Investigate critically. Present in detail and discuss the implications critically.

explain Answer "how" and "why" questions. Clarify, using concrete examples.

footnote Short note placed at the bottom of a page and identified by a symbol or numeral in the body of the text. A textual "aside," a footnote provides a brief elaboration of some point made in the text whose inclusion there might disrupt the flow of text. In the *note system* of referencing, footnotes may also include details of reference material cited in the text.

free writing "Stream of consciousness" writing without concern for overall structure and direction, followed by careful revision. Sometimes used as a step in the production of an essay.

generalization Comprehensive statement about all or most examples of some phenomenon made on the basis of a (limited) number of observations of examples of that phenomenon.

Harvard system See *author-date system*.

histogram Graph in which plotted values are shown in the form of horizontal or, more commonly, vertical bars whose area is proportional to the value(s) portrayed. Thus, if class intervals depicted in the histogram are of different sizes, the column areas will reflect this. (Compare with *bar graph*.)

home page "Front" page of a website or the "top" page in a hierarchy of web pages. Usually sets out the website's content and other characteristics. (See also *web page* and *website*.)

hypothesis Supposition or trial proposition used as a starting point for investigation. (e.g., "My hypothesis is that for distances greater than 500 kilometres, transporting goods by rail is a more efficient use of energy than transporting goods by truck.")

ibid. Abbreviation of the Latin word *ibidem*, meaning "in the same place"; sometimes used in footnotes and endnotes.

illustrate Make clear through the use of examples or by use of figures, diagrams, maps, and photographs.

indicate Focus attention on or point out.

Internet Worldwide network of computer networks that allows communication between computers connected to the network.

interpret Make clear, giving your own judgment; offer an opinion or reason for the character of some phenomenon.

introduction First section in a piece of formal communication (e.g., poster, talk, essay), in which author or speaker tells the audience what is going to be discussed and why.

invigilator Person who administers a test or examination and who may be different from the person who teaches the course.

isoline Line connecting points of known, or estimated, equal values.

jargon Most commonly, technical terms used unnecessarily when clearer terms would suffice. Less commonly, words or a mode of language intelligible only to a group of experts in the field.

journal Publication consisting of articles, issued, at regular or irregular intervals, on an ongoing basis (e.g., *The Canadian Geographer, Ecoscience*). See *periodical*.

justify Provide support and evidence for outcomes or conclusions.

key See *legend*.

legend Brief explanatory statement on a map or diagram about the symbols, patterns, and colours used in a map or diagram. Also known as a *key*.

line graph Graph in which the values of observed phenomena (represented by *x* and *y*) are connected by lines. Used to illustrate change over time or relationships between variables.

listserver Software program that automatically distributes email messages to names on a mailing list.

literature review Comprehensive summary and interpretation of written resources (e.g., books, reports, articles) and their relationship to a specific area of research.

loc. cit. Abbreviation of the Latin phrase *loco citato*, meaning "in the place cited"; occasionally used in footnotes and endnotes.

logarithmic scale Scale on a graph axis in which the key intervals are based on exponents of 10 (for a base-10 scale). The base value can be different, but 10 is the most commonly used base for logarithmic scales. (Compare with *arithmetic scale*.)

log-log graph Graph with logarithmic scales on both axes. (Compare with *semi-log graph*.)

media release Information pertaining to an issue or event sent to a media outlet. Also known as a "press release."

note identifier Symbol or numeral used in text to refer a reader to a reference or to supplementary information in an endnote or footnote.

note system System of referring to texts cited. Comprises a superscript numeral or other symbol in the text that refers the reader to a footnote or endnote containing the full bibliographic details of the reference. Compare with *author-date system*.

op. cit. Abbreviations of the Latin phrase *opere citato*, meaning "in the work cited"; used occasionally in footnotes and endnotes.

oral presentation Delivering an address to an audience orally. Often accompanied by supporting *visual aids*.

orthophoto map Map created from a mosaic of aerial photographs or satellite images. May be overlain with information, such as gridlines, contours, transport routes, and place names.

outline *noun*: Brief sketch or written plan. *verb*: To describe the main features, leaving out minor details.

paragraph Cohesive, self-contained expression of an idea, usually constituting part of a longer written document. Usually has three parts: *topic sentence*, *supporting sentence(s)*, and *concluding sentence*.

paraphrasing Summarizing someone else's words in your own.

parentheses Punctuation, written as a pair of round brackets (such as these), placed around a word or group of words that are inserted into a sentence by way of explanation but that are not grammatically necessary to the sentence.

peer review Process by which a piece of work is examined carefully by a person or persons with expertise in the field of inquiry before the work is published.

periodical Magazine or journal published at regular intervals.

pie graph Circular graph in which proportions of some total sum (the whole "pie") are depicted as "slices." The area of each slice is directly proportional to the size of the variable portrayed.

plagiarism Act of presenting, without proper attribution, someone else's work or ideas as one's own.

population pyramid Form of histogram showing the number or percentage of people in different age groups of a population.

poster Graphic display made of some kind of stiff or rigid material to which are affixed textual and graphic materials outlining the results of some piece of research. Should minimize amount of text in favour of showing illustrative materials that support the argument being made or story being told.

PowerPoint Proprietary software (from Microsoft Corp.) used in public presentation of text or graphics. Includes tools for word processing, outlining, drawing, graphing, and presentation management.

précis See *summary*.

preface Section at the beginning of a book or report in which the author briefly states how the book came to be written and what its purpose is. May include acknowledgements unless they are given in a separate section.

prove Demonstrate truth or falsity by use of evidence.

qualitative methods Collection of data, often with a degree of subjectivity, by observing what people think and say, and complementary techniques of analysis and interpretation. (Compare with *quantitative methods*.)

quantitative methods Collection, analysis, and interpretation of data using objective techniques. (Compare with *qualitative methods*.)

quotation (or quote) Verbatim (i.e., word-for-word) reproduction of someone else's words.

quotation marks Marks used to indicate that words have been reproduced exactly from someone else's speech or writing. (e.g., Albert Einstein wrote, "The most beautiful experience we can have is the mysterious.") Quotation marks may also be used to draw the reader's attention to a word that is used in an unusual or incorrect way, but this device should not be overused.

reference Source referred to or used in the preparation of a formal communication (e.g., essay, book review, poster, report).

reference list Complete list of works actually cited or acknowledged in a formal communication. (Compare with *bibliography*.)

relate Establish and show the connections between one phenomenon and another.

reference style Set of formatting rules for citing sources and preparing entries in a reference list.

replication Repetition of a previous experiment or study for the purpose of discovering whether or not the same results will be obtained, thereby lending weight (or not) to the results of the original study or experiment. (Compare with *corroboration*.)

rhetorical question Question that is asked to stimulate thought or make a point, but where no answer is expected to be given.

scale Indication on a map or diagram of the relationship between the size of some depicted phenomenon and its size in reality. (See also *cartographic scale*.)

scatter plot Graph of point data plotted by their (x, y) coordinates.

self-publishing Dissemination to the public of a book, article, report, etc., by the author rather than by a professional publisher. May be done on the Internet as well as in traditional print form.

semi-log graph Graph on which one axis has a logarithmic scale and one has a numerical scale.

sentence Group of words that expresses a complete thought and contains a subject and a finite verb.

sic Latin word that means "thus." It is used with a direct quotation containing an error or a questionable statement. Usually italicized and enclosed in square brackets: [*sic*]. It is used to mean "this is the way it appeared." If only one word is wrong in the quotation, [*sic*] appears directly after that word: "He done [*sic*] great work on the field today." Otherwise, it follows the quotation.

source Material containing facts, ideas, or data that are used as evidence to support assertions or arguments, or to provide context for review or discussion of a topic.

spam Email message sent in bulk form by an unknown sender to a large number of recipients.

state Express fully and clearly.

summarize Present critical points in brief, clear form.

summary Brief description of a piece of writing or a talk.

supplementary (or illustrative) materials Items other than written text that support or enrich the work. Examples include diagrams, maps, graphs, tables, charts, and photographs.

supporting sentence Sentence in the part of a paragraph that presents a discussion substantiating the paragraph's claim(s). (See also *topic sentence* and *concluding sentence*.)

synthesize Build up separate elements into some comprehensible whole. (Compare with *analyze*.)

table Systematically arranged list of facts or numbers, usually arranged in rows and columns.

title Distinguishing name or description of a communication. Should be kept short but also contain enough specific information to be representative of the communication's contents.

topic sentence Sentence in a paragraph in which the main idea is expressed. (See also *supporting sentence* and *concluding sentence*.)

topographic map Common, general-reference map that usually depicts contours, physical features (e.g., rivers, peaks), and cultural features (e.g., roads, churches, cemeteries).

trace Describe the development of a phenomenon from some origin(s).

URL Stands for Uniform Resource Locator, a string of letters and numbers that makes up the address of a website.

visual aids Materials used in an oral presentation to support content being delivered verbally.

web See *World Wide Web*.

web page Computer file with a unique web location that may be grouped with other related web pages to form a *website*.

website Collection of related and linked *web pages*.

World Wide Web International computer network consisting of all sites on the Internet adhering to HTTP address conventions.

References

Academic Skills Centre. (1995). *Thinking it through: A practical guide to academic essay writing* (revised 2nd ed.). Peterborough, ON: Trent University.

Alley, M. (2003). *The craft of scientific presentations*. New York: Springer Science+Business Media.

———. (undated). *Design of scientific posters*. http://writing.engr.psu.edu/posters.html (accessed 18 November 2013).

Anglia Ruskin University. (undated). *Harvard system of referencing guide*. http://libweb.anglia.ac.uk/referencing/harvard.htm (accessed 18 November 2013).

Anholt, R.R. (2006). *Dazzle 'em with style: The art of oral scientific presentation* (2nd ed.). Burlington, MA: Elsevier.

Arctic Power—Arctic National Wildlife Refuge. (2013). Home page. www.anwr.org (accessed 20 November 2013).

Australian Bureau of Statistics. (1994). *Australia Yearbook*, Cat. No. 1301. Canberra: AGPS.

Barass, R. (1984). *Study! A guide to effective study, revision and examination techniques*. London: Chapman and Hall.

Barzun, J., and Graff, H.F. (2004). *The modern researcher* (6th ed.). Belmont, CA: Thomson Wadsworth.

Bate, D., and Sharpe, P. (1990). *Student writer's handbook*. Marrickville, Australia: Harcourt Brace Jovanovich.

BC Lung Association. (2009). *State of the air 2009: British Columbia*. www.bc.lung.ca/airquality/documents/SOA2009webmedres.pdf (accessed 25 January 2010).

Beck, S.E. (2009). *Evaluation Criteria*. http://lib.nmsu.edu/instruction/evalcrit.html (accessed 20 November 2013).

Beins, B.C., and Beins, A.M. (2008). *Effective writing in psychology*. Malden, MA: Blackwell.

Beynon, R.J. (1993). *Postgraduate study in the biological sciences: A researcher's companion*. London: Portland Press.

Booth, V. (1993). *Communicating in science: Writing a scientific paper and speaking at scientific meetings* (2nd ed.). Cambridge: Cambridge University Press.

Bowman, V. (2004a). Teaching intellectual honesty in a tragically hip world: A pop-culture perspective. In V. Bowman (ed.), *The plagiarism plague*. New York: Neal-Schuman, pp. 3–11.

——— (ed.). (2004b). *The plagiarism plague*. New York: Neal-Schuman.

Boyd, B., and Taffs, K. (2003). *Mapping the environment: A professional development manual*. Lismore, Australia: Southern Cross University Press.

BP. (2004) *Energy in focus: BP statistical review of world energy, June 2004*. www.bp.com/liveassets/bp_internet/globalbp/STAGING/global_assets/downloads/S/statistical_review_of_world_energy_full_report_2004.pdf (accessed 22 November 2013).

Breckenridge, D. (2008). *PR 2.0: New media, new tools, new audiences*. Upper Saddle River, NJ: FT Press.

Bregha, F. (2012). *National Energy Program*. www.thecanadianencyclopedia.com/index.cfm?PgNm=TCE&Params=A1ARTA0005618 (accessed 20 November 2013).

Brozio-Andrews, A. (2007). *Writing bad-book reviews*. www.absolutewrite.com/specialty_writing/bad_book_reviews.htm (accessed 18 November 2013).

Buckley, J. (2009). *Fit to print: The Canadian student's guide to essay writing*. Toronto: Nelson Education.

Burdess, N. (1998). *The handbook of student skills for the social sciences and humanities* (2nd ed.). New York: Prentice Hall.

Burkill, S., and Abbey, C. (2004). Avoiding plagiarism. *Journal of Geography in Higher Education, 28*(3), 439–446.

Callicott, J.B. and Froderman, R. (eds.). (2008). *Encyclopedia of environmental ethics and philosophy*, vol. 2. Detroit: Gale Cengage Learning.

Canada Mortgage and Housing Corporation. (2009). *CHS—Residential building activity, dwelling starts, completions, under construction and newly completed and unabsorbed dwellings—2008*. www.cmhc-schl.gc.ca/odpub/esub/64681/64681_2009_A01.pdf (accessed 27 October 2009).

Canadian Encyclopedia. (2011). *About the Canadian Encyclopedia*. www.thecanadianencyclopedia.com/index.cfm?PgNm=AboutThisSite&Params=A1 (accessed 20 November 2013).

Canadian Journal of Earth Sciences. (2013). *Instructions to authors*. www.nrcresearchpress.com/page/cjes/authors (accessed 21 November 2013).

Cargill, M., and O'Connor, P. (2009). *Writing scientific research articles: Strategy and steps*. Chichester, UK: Wiley-Blackwell.

Carney, W.W. (2008). *In the news: The practice of media relations in Canada* (2nd ed.). Edmonton, AB: University of Alberta Press.

Cartography Specialty Group of the Association of American Geographers. (1995). Guidelines for effective visuals at professional meetings. *AAG Newsletter*, *50*(7), 5.

Cartwright, W., Peterson, M.P., and Gartner, G. (eds.). (2007). *Multimedia cartography*. Berlin: Springer.

Central Intelligence Agency. (undated). *The world factbook*. www.cia.gov/library/publications/the-world-factbook/index.html (accessed 20 November 2013).

Central Queensland University Library. (2000). *Why do a literature review?* www.library.cqu.edu.au/litreviewpages/why.htm (accessed 6 October 2009).

Centre for Topographic Information. (2007). *The national topographic system of Canada*. www.nrcan.gc.ca/earth-sciences/geography-boundary/mapping/topographic-mapping/10339 (accessed 20 November 2013).

Clanchy, J., and Ballard, B. (1997). *Essay writing for students: A practical guide* (3rd ed.). Melbourne: Longman Cheshire.

Cottingham, C., Healey, M., and Gravestock, P. (2002). *Fieldwork in the geography, earth and environmental sciences higher education curriculum*. www2.glos.ac.uk/gdn/disabil/fieldwk.htm (accessed 19 November 2013).

Cottrell, S. (2003). *The study skills handbook* (2nd ed.). Basingstoke, UK: Palgrave Macmillan.

Courtenay, B. (1992). *The pitch*. McMahons Point, Australia: Margaret Gee.

Cvetkovic, N. (2004). The dark side of the web: Where to go to buy a paper. In V. Bowman (ed.), *The plagiarism plague*. New York: Neal-Schuman, pp. 25–34.

Deakin University Library. (undated). *The literature review*. www.deakin.edu.au/library/findout/research/litrev.php (accessed 21 November 2013).

DeLyser, D., and Pawson, E. (2005). From personal to public: Communicating qualitative research for public consumption. In I. Hay (ed.), *Qualitative research methods in human geography* (2nd ed.). Melbourne: Oxford University Press, pp. 266–274.

Dixon, T. (2004). *How to get a first: The essential guide to academic success*. London: Routledge.

Driscoll, D.L., and Brizee, A. (2012). *Appropriate language: Overview*. https://owl.english.purdue.edu/owl/resource/608/1/ (accessed 21 November 2013).

Eisenberg, A. (1992). *Effective technical communication* (2nd ed.). New York: McGraw-Hill.

Eisner, C., and Vicinus, M. (eds.). (2008). *Originality, imitation and plagiarism*. Ann Arbor, MI: University of Michigan Press.

Elections Canada. (2013). *Home page*. www.elections.ca/home.asp (accessed 20 November 2013).

Engle, M. (2013). *How to prepare an annotated bibliography*. Cornell University. www.library.cornell.edu/olinuris/ref/research/skill28.htm (accessed 18 November 2013).

Environment Canada. (2013). *Climate*. http://climate.weather.gc.ca/index_e.html (accessed 21 November 2013).

Estonian Tourist Board. (undated). *Estonian Flag Room*. www.visitestonia.com/en/estonian-flag-room (accessed 20 November 2013).

Foremski, T. (2006). *Die! Press release! Die! Die! Die!* www.siliconvalleywatcher.com/mt/archives/2006/02/die_press_relea.php (accessed 18 November 2013).

FreeAlberta.com. (undated). *National Energy Program*. www.freealberta.com/nep.html (accessed 20 November 2013).

Friedman, S.F., and Steinberg, S. (1989). *Writing and thinking in the social sciences*. New York: Prentice-Hall.

Game, A., and Metcalfe, A. (2003). *The first year experience: Start, stay and succeed at uni*. Leichhardt, Australia: Federation Press.

Glaser, M. (2006). *Your guide to citizen journalism*. www.pbs.org/mediashift/2006/09/your-guide-to-citizen-journalism270.html (accessed October 6, 2009).

Goldbort, R. (2006). *Writing for science*. New Haven, CT: Yale University Press.

Golledge, R.G., and Garling, T. (2004). Cognitive maps and urban travel. In D.A. Hensher, K.J. Button, K.E. Haynes, and P.R. Stopher (eds.), *Handbook of transport geography and spatial systems*. Amsterdam: Elsevier, pp. 501–512.

Gould, P. (1993). *The slow plague: A geography of the AIDS pandemic*. Cambridge, MA: Blackwell.

Grantham, D.W. (1987). *Recent America*. Arlington Heights, IL: Harlan Davidson.

Greetham, B. (2008). *How to write better essays*. Basingstoke, UK: Palgrave.

Gurak, L. (2000). *Oral presentations for technical communication*. Boston: Allyn and Bacon.

Gustavii, B. (2008). *How to write and illustrate a scientific paper* (2nd ed.). New York: Cambridge University Press.

Hamilton, D. (1999). *Passing exams: A guide for maximum success and minimum stress*. London: Cassell.

Harris, R.A. (2011). *Using sources effectively: Strengthening your writing and avoiding plagiarism* (3rd ed.). Los Angeles, CA: Pyrczak.

Hay, I. (1994). Justifying and applying oral presentations in geographical education. *Journal of Geography in Higher Education, 18*(1), 43–55.

Hay, I., and Thomas, S. (1999). Making sense with posters in biological science education. *Journal of Biological Education, 33*(4), 209–214.

Hay, I., Dunn, K., and Street, A. (2005). Making the most of your conference journey. *Journal of Geography in Higher Education, 29*(1), 159–171.

Hess, G.R., Tosney, K.W., and Liegel, L.H. (2013). *Creating effective poster presentations*. www.ncsu.edu/project/posters (accessed 19 November 2013).

Hillstrom, K. (2005). *Defining moments: The Internet revolution*. Detroit, MI: Omnigraphics.

Hodge, D. (1994a). *Writing a good term paper*. Course handout, Department of Geography, University of Washington, Seattle.

———. (1994b). *Guidelines for professional reports*. Course handout for GEOG 426, Department of Geography, University of Washington, Seattle.

Hovius, N., Lague, D., and Dadson, S. (2004). Processes, rates and patterns of mountain-belt erosion. In P.N. Owens and O. Slaymaker, *Mountain geomorphology*. London: Arnold.

Howitt, R., Connell, J., and Hirsch, P. (eds.). (1996). *Resources, nations and indigenous peoples: Case studies from Australasia, Melanesia and Southeast Asia*. Melbourne: Oxford University Press.

ICMJE (International Committee of Medical Journal Editors). (2013). *Uniform requirements for manuscripts submitted to biomedical journals: Writing and editing for biomedical publication*. www.icmje.org/urm_full.pdf (accessed 17 November 2013).

Jeffries, D.S., Brydges, T.G., Dillon, P.J., and Keller, W. (2003). Monitoring the results of Canada/U.S.A. acid rain control programs: Some lake responses. *Environmental Monitoring and Assessment, 88*(1–3), 3–19.

Jenkins, A., and Pepper, D. (1988). Enhancing students' employability and self-expression: How to teach oral and groupwork skills in geography. *Journal of Geography in Higher Education, 12*(1), 67–83.

Jennings, J.T. (1990). *Guidelines for the preparation of written work* (4th ed.). Roseworthy, Australia: University of Adelaide, Roseworthy Campus.

Johnson, J. (1997). *The Bedford guide to the research process*. Boston, MA: Bedford/St Martins.

Kimerling, A.J., Muehrcke, P.C., and Muehrcke, J.O. (2005). *Map use: Reading, analysis, and interpretation*. Madison, WI: JP.

Kneale, P. (1997). Maximising play time: Time management for geography students. *Journal of Geography in Higher Education, 21*(2), 293–301.

Knight, P., and Parsons, T. (2003). *How to do your essays, exams, and coursework in geography and related disciplines*. Cheltenham, UK: Nelson Thornes.

Kosslyn, S.M. (2006). *Graph design for the eye and mind*. New York: Oxford University Press.

Krohn, J. (1991). Why are graphs so central in science? *Biology and Philosophy, 6*(2), 181–203.

Krygier, J., and Wood, D. (2005). *Making maps: A visual guide to map design for GIS*. New York: Guilford Press.

Lethbridge, R. (1991). *Techniques for successful seminars and poster presentations*. Melbourne: Longman Cheshire.

Ley, D.F., and Bourne, L.S. (1993). Introduction: The social context and diversity of urban Canada. In L.S. Bourne and D.F. Ley (eds.), *The changing social geography of Canadian cities*. Montreal: McGill-Queen's University Press.

Library of Congress. (undated). *How to cite electronic sources*. www.loc.gov/teachers/usingprimarysources/citing.html (accessed 27 November 2009).

Lindholm-Romantschuk, Y. (1998). *Scholarly book reviewing in the social sciences and humanities*. Westport, CT: Greenwood Press.

Lovell, D.W, and Moore, R.D. (1992). *Essay writing and style guide for politics and the social sciences*. Canberra: Australasian Political Studies Association.

MacEachren, A.M. (1995). *How maps work*. New York: Guilford Press.

Mansvelt, J., and Berg, L. (2005). Writing qualitative geographies, constructing geographical knowledges. In I. Hay (ed.), *Qualitative research methods in human geography* (2nd ed.). Melbourne: Oxford University Press, pp. 248–265.

Marius, R., and Page, M.E. (2002). *A short guide to writing about history*. New York: Longman.

May, C.A. (2007). *Spotlight on critical skills in essay writing*. Toronto: Pearson Education Canada.

Mills, C. (1994). Acknowledging sources in written assignments. *Journal of Geography in Higher Education, 18*(2), 263–268.

Mohan, T., McGregor, H., and Strano, Z. (1992). *Communicating! Theory and practice* (3rd ed.). Sydney: Harcourt Brace.

Monmonier, M. (1996). *How to lie with maps* (2nd ed.). Chicago: University of Chicago Press.

Montello, D.R., and Sutton, P.C. (2006). *An introduction to scientific research methods in geography*. Thousand Oaks, CA: Sage.

Montgomery, D.R. (2001). Slope distributions, threshold hillslopes, and steady-state topography. *American Journal of Science, 301*(4–5), 432–454.

Montgomery, S.L. (2003). *The Chicago guide to communicating science*. Chicago: University of Chicago Press.

Morgan, S., and Whitener, B. (2006). *Speaking about science: A manual for creating clear presentations*. New York: Cambridge University Press.

Moxley, J.M. (1992). *Publish, don't perish: The scholar's guide to academic writing and publishing*. Westport, CT: Praeger.

Najar, R., and Riley, L. (2004). *Developing academic writing skills*. Tokyo: Macmillan Language house.

National Foresty Database. (2009). *Silviculture—quick facts*. www.nfdp.ccfm.org/silviculture/quick_facts_e.php (accessed 27 October 2009).

National Geographic Society. (undated). *National Geographic Map Collection*. www.ngmapcollection.com/ (accessed 20 November 2013).

Natural Resources Canada. (2009). *The Atlas of Canada—Land and freshwater areas*. http://atlas.nrcan.gc.ca/site/english/learningresources/facts/surfareas.html (accessed 27 October 2009).

New Zealand Tourism Board. (1995). *New Zealand where to stay guide*. Wellington: New Zealand Tourism Board.

Nick, J.M. (2012). Open access part 1: The movement, the issues, and the benefits. The *Online Journal of Issues in Nursing, 17*(1), unpaginated. www.nursingworld.org/MainMenuCategories/ANAMarketplace/ANAPeriodicals/OJIN/TableofContents/Vol-17-2012/No1-Jan-2012/Articles-Previous-Topics/Open-Access-Part-I.html (accessed 22 February 2014).

Nicol, A.A.M., and Pexman, P.M. (2003). *Displaying your findings: A practical guide for creating figures, posters, and presentations*. Washington, DC: American Psychological Association.

Northey, M., Knight, D.B., and Draper, D. (2012). *Making Sense: Geography and Environmental Sciences* (5th ed.). Toronto: Oxford University Press.

Oliver, M. (2000). *Using AutoCAD Map 2000*. New York: Autodesk Press.

Orr, F. (2005). *How to pass exams*. Sydney: George Allen and Unwin.

Pechenik, J.A. (2007). *A short guide to writing about biology* (6th ed.). New York: Pearson Longman.

Perrella, A.M.L. (2009). Economic decline and voter discontent. *Social Science Journal, 46*, 347–368.

Peterson, G.N. (2009). *GIS cartography: A guide to effective map design*. Boca Raton, FL: CRC Press.

Peterson, M.P. (2008). *International perspectives on maps and the Internet*. Berlin: Springer.

Poljak, Z., Friendship, R.M., Carman, S., McNab, W.B., and Dewey, C.E. (2008). Investigation of exposure to swine influenza viruses in Ontario (Canada) finisher herds in 2004 and 2005. *Preventive Veterinary Medicine, 83*, 24–40.

Preston, N. (2001). *Understanding ethics* (2nd ed.). Sydney: Federation Press.

Purrington, C.B. (undated). *Advice on designing scientific posters*. www.swarthmore.edu/NatSci/cpurrin1/posteradvice.htm (accessed 19 November 2013).

Radford, M.L., Barnes, S.B., and Barr, L.R. (2002). *Web research: Selecting, evaluating, and citing*. Boston, MA: Allyn and Bacon.

Reece, P. (1999). The number one fear: Public speaking and the university student. In K. Martin, N. Stanley, and N. Davison (eds.), *Teaching in the disciplines/learning in context*. Proceedings of the 8th Annual Teaching Learning Forum, University of Western Australia. http://otl.curtin.edu.au/tlf/tlf1999/reece.html (accessed 26 October 2009).

Reed, T.V. (2011). *Environmental justice cultural studies bibliography*. http://culturalpolitics.net/environmental_justice/bibliography (accessed 30 May 2011).

Robinson, A.H., Morrison, J.L., Muehrcke, P.C., Guptill, S.C., and Kimerling, A.J. (1994). *Elements of cartography* (6th ed.). New York: John Wiley.

Slocum, T.A, McMaster, R.B., Kessler, F.C., and Howard, H.H. (2009). *Thematic cartography and geovisualization* (3rd ed.). Upper Saddle River, NJ: Pearson Education.

Statistics Canada. (2009). Home page. www.statcan.gc.ca/start-debut-eng.html (accessed 20 November 2009).

Stettner, M. (1992). How to speak so facts come to life. In D.F. Beer (ed.), *Writing and speaking in the technology professions: A practical guide*. New York: IEEE, pp. 225–228.

Thody, A. (2006). *Writing and presenting research*. London: Sage.

Tracy, E. (2002). *The student's guide to exam success*. Buckingham, UK: Open University Press.

Translation Bureau. (1997). *The Canadian style: A guide to writing and editing*. Toronto: Dundurn Press.

Transport for London. (undated). *Tube map*. www.tfl.gov.uk/assets/downloads/standard-tube-map.pdf (accessed 20 November 2013).

Truss, L. (2004). *Eats, Shoots & Leaves: The Zero Tolerance Approach to Punctuation*. New York: Gotham Books.

Turabian, K. (2013). *A manual for writers of research papers, theses, and dissertations* (8th ed. revised by Booth, W.C., Colomb, G.G., Williams, J.M., and the University of Chicago Press editorial staff). Chicago: University of Chicago Press.

U.S. Census Bureau. (2009). *Historical estimates of world population*. www.census.gov/ipc/www/worldhis.html (accessed 27 October 2009).

U.S. Department of Agriculture, Economic Research Service. (2009). *International macroeconomic data set*. www.ers.usda.gov/data/macroeconomics/ (accessed 27 October 2009).

U.S. Energy Information Administration. (2013). Home page. www.eia.gov/ (accessed 22 November 2013).

UC Berkeley. (2012). *Evaluating web pages: Techniques to apply and questions to ask*. www.lib.berkeley.edu/TeachingLib/Guides/Internet/Evaluate.html (accessed 18 November 2013).

United Nations Population Division. (2012). *World population prospects: The 2012 revision*. New York: United Nations Department of Economic and Social Affairs, Population Division. http://esa.un.org/unpd/wpp/Excel-Data/population.htm (accessed 18 November 2013).

University of California, Santa Cruz University Library. (undated). *Write a literature review*. http://library.ucsc.edu/ref/howto/literaturereview.html (accessed 21 November 2013).

University of Texas Libraries. (2009). *Perry-Casteñada Map Collection—Thematic Maps*. http://lib.utexas.edu/maps/thematic.html (accessed 20 November 2009).

University of Wisconsin, Madison Writing Centre. (2012). *Review of literature*. www.wisc.edu/writing/Handbook/ReviewofLiterature.html (accessed 18 November 2013).

UO Libraries. (undated). *Looking for articles in journals and magazines: Scholarly or popular?* http://library.uoregon.edu/guides/findarticles/distinguish.html (accessed 21 November 2013).

Van Blerkom, D.L. (2009). *College study skills: Becoming a strategic learner* (6th ed.). Boston, MA: Wadsworth Cengage Learning.

Vicente, M.A. (undated). *The National Energy Program*. www.abheritage.ca/abpolitics/events/issues_nep.html (accessed 20 November 2013).

Vujakovic, P. (1995). Making posters. *Journal of Geography in Higher Education, 19*(2), 251–256.

Wainer, H. (1984). How to display data badly. *The American Statistician, 38*(2), 137–147.

Waitt, G., McGuirk, P., Dunn, K., Hartig, K., and Burnley, I. (2000). *Introducing human geography*. French's Forest, Australia: Longman.

Walters, T.N., Walters, L.M., and Starr, D.P. (1994). After the highwayman: Syntax and successful placement of press releases in newspapers. *Public Relations Review, 20*(4), 345–356.

Weintraub, I. (1994). Fighting environmental racism: A selected annotated bibliography. *Electronic Green Journal, 1*(1), unpaginated.

Weston, L., and Handy, S. (2004). Mental maps. In D.A. Hensher, K.J. Button, K.E. Haynes, and P.R. Stopher (eds.), *Handbook of transport geography and spatial systems.*, Amsterdam: Elsevier, pp. 533–545.

Wikipedia. (2013a). *About Wikipedia*. http://en.wikipedia.org/wiki/Wikipedia:About (accessed 20 November 2013).

Wikipedia. (2013b). National Energy Program. http://en.wikipedia.org/wiki/National_Energy_Program (accessed 20 November 2013).

Williams, D. (1994). In defense of the (properly executed) press release. *Public Relations Quarterly*, Fall, 5–7.

Windschuttle, K., and Elliott, E. (1999). *Writing, researching, communicating* (3rd ed.). Sydney: McGraw-Hill.

Woodford, F.P. (1967). Sounder thinking through clearer writing. *Science, 156* (3776), 744.

World Heritage Centre. (2013). *World Heritage List statistics*. Paris: UNESCO World Heritage Centre. http://whc.unesco.org/en/list/stat#s6 (accessed 20 November 2013).

Writing Center, University of North Carolina at Chapel Hill. (2013). *Book reviews*. http://writingcenter.unc.edu/handouts/book-reviews/ (accessed 19 November 2013).

Writing Center, Wilkes University. (undated). *Sexist and biased language*. www.wilkes.edu/pages/776.asp (accessed 19 November 2013).

Zittrain, J. (2008). *The future of the Internet— and how to stop it*. New Haven, CT: Yale University Press.

Credits

Grateful acknowledgement is made for permission to reprint the following:

Table 2.2: Adapted from Beck (2009) and UC Berkeley (2012).

Box 5.1, Example 2: Marketing text for *Environmentalism in Popular Culture* by Noël Sturgeon © 2008 The Arizona Board of Regents. Reprinted by permission of the University of Arizona Press.

Box 5.1, Example 3: *Fighting Enviromental Racism: A Selected Annotated Bibliography*, by Irwin Weintraub, Electronic Green Journal, Issue 1, June, 1994

Box 5.5, Examples 1, 2 and 3: Reproduced with permission from *Library Journal*. Copyright Library Journals, LLC, a wholly owned subsidiary of Media Source Inc.

Fig. 7.1: Hay and Thomas, 'Making Sense with Posters in Biological Science Education', *Journal of Biological Education*, vol. 33, no. 4 (1999, p. 211).

Figure 8.1: United Nations Population Division (2012).

Figure 8.2: Data Source: World Heritage Centre (2013).

Figure 8.3: BC Lung Association (2009).

Figure 8.7: Jeffries, D.S., Brydges, T.G., Dillon, P.J., and Keller, W. (2003). Monitoring the results of Canada/U.S.A. acid rain control programs: Some lake responses. *Environmental Monitoring and Assessment*, 88(1–3), 3–19.

Figure 8.8: CHS: Residential Building Activity—Dwelling Starts, Completions, Under Construction and Newly Completed and Unabsorbed Dwellings, Canada Mortgage and Housing Corporation (2009).

Figure 8.9: Source: Natural Resources Canada.

Figure 8.10: Statistics Canada (2009). Reproduced and distributed on an "as is" basis with the permission of Statistics Canada.

Figure 8.11: Statistics Canada (2009). Reproduced and distributed on an "as is" basis with the permission of Statistics Canada.

Figure 8.12: Statistics Canada (2009). Reproduced and distributed on an "as is" basis with the permission of Statistics Canada.

Figure 8.13: United Nations Population Division (2012).

Figure 8.14: Environment Canada (2013).

Figure 8.15: Statistics Canada (2009). Reproduced and distributed on an "as is" basis with the permission of Statistics Canada.

Figure 8.16: Statistics Canada (2009). Reproduced and distributed on an "as is" basis with the permission of Statistics Canada.

Figure 8.17: US Census Bureau (2009).

Figure 8.18: US Department of Agriculture, Economic Research Services (2009).

Table 8.5: Canadian Council of Forest Ministers, National Forestry Database (http://nfdp.ccfm.org). Reproduced with permission of Natural Resources Canada. 2014.

Figure 9.1: Poljak et al (2008).

Figure 9.2: Statistics Canada (2009). Reproduced and distributed on an "as is" basis with the permission of Statistics Canada.

Figure 9.5: Statistics Canada (2009). Reproduced and distributed on an "as is" basis with the permission of Statistics Canada.

Figure 9.6: Adapted from: University of Texas Libraries. (2009). Perry-Casteñada Map Collection—Thematic Maps. http://lib.utexas.edu/maps/thematic.html (accessed 20 November 2009).

Figure 9.7: Adapted from: http://www.resilience.org/stories/2007-11-17/who-has-oil (accessed 23 July 2014). Data source: BP (2004) and U.S. Energy Information Administration (2013).

Box 11.5: SMU public affairs. Used with permission.

Index

abbreviations, 71, 128
abstracts: descriptive, 68, 70; informative, 68, 69–70; poster, 149; report, 68–71
academic essay, 47
Academic Skills Centre, 60
acknowledgements: poster, 150; reasons for, 35; sources and, 33–5; report, 71
acronyms, 71
aggregated data: maps and, 195
Alberta Online Encyclopedia, 32–3
Alley, M., 144, 157, 232
ampersand, 117
Anglia Ruskin University, 142
Anholt, R.R., 232
annotated bibliography, 88–91; purpose of, 90
APA (American Psychological Association) style, 112, 120
apostrophe, 140–1
appendices, 15, 79
ArcGIS, 205
Arctic Power, 28
argument, 47; essays and, 48–56
argumentative essays, 47
assessment: posters and, 145–8
assessment forms: essay, 58–60; oral presentation, 211–13; poster, 147–8; report, 81–3; review, 97–9; summary/précis, 93–4
Atwan, Robert, 46
audience: oral presentations and, 214, 227–30; posters and, 146, 150–1; public, 234–51; questions from, 230–1; visual aids and, 223; writing and, 4–5
author–date system, 110, 112, 114–27
authors: credibility and, 28; group, 118, 121–2; multiple, 117–18; multiple works by, 127; review and, 99; unknown, 118, 129; websites and, 32
AutoCAD, 205

Bacon, Francis, 88
Baines, P., et al., 250
bar graphs, 163, 169–73; horizontal/vertical, 169
Barass, R., 269
Barzun, J., and Graff, H.F., 21, 45
Bate, D., and Sharpe, P., 9
Bates, Marston, 62
Beck, Harry, 198
Beck, S.E., 45
Beins, B.C., and Beins, A.M., 157–8
bias: language and, 134–6; sexual, 135
bibliographic details, 12, 73, 110, 111; annotated bibliography and, 91; plagiarism and, 40–1; summary and, 92
bibliography, 13, 88, 134; annotated, 88–91; note system and, 128
blackboard, 223
Bland, M., et al., 250
blogs, 30, 31, 236; self-publishing and, 247–50
body: essay, 51, 53–4; oral presentation, 218; poster, 149–50
Bonner, A., 250
books, 22, 23–4; electronic, 125, 132–3; in note system, 128–9; in reference list, 120
Booth, V., 86, 232
border, map, 206
Bowman, L., 232
Bowman, V., 42, 45
Breckenridge, D., 239, 250
Brown, B.S., 158
Brozio-Andrews, A., 107
Bryan, William Jennings, 209
Buckley, J., 60
Burdess, N., 107, 269
Burkill, S., and Abbey, C., 40, 42

Calnan, J., and Barabas, A., 232
Canadian Association of Geographers, 248
Canadian Encyclopedia, 32–3
captions, 15, 156
Cargill, M., and O'Connor, P., 67
Carle, G., 232
Carney, W.W., 237, 249, 251
cartograms, 198
cartographic scale, 201–2
CD-ROM: in note system, 131; in reference list, 123
Centre for Topographic Information, 200–1
chapters: in note system, 129; in reference list, 121
charts, 14; *see also* graphs
choropleth maps, 195–7
citations, 12–13, 109–34; in-text, 114–19; in note system, 128; report, 73
Clanchy, J., and Ballard, B., 269
class intervals, 174, 176–8, 196
closed-book examinations, 253
colon, 139–40
colour: maps and, 208; posters and, 153–4
Colton, Charles Caleb, 252
comma, 136, 137–8; citations and, 116
comments: essays drafts and, 4, 5
common knowledge, 33, 38–9
computers: writing and, 9; *see also* software
concept map, 258
conceptual framework: essay and, 53; review and, 99–100

conclusions: essay, 51, 53–4; oral presentation, 218–19; poster, 150; report, 77–8
conference papers: in note system, 129–30; in reference list, 121
conferences: public speaking and, 210
consistency: reference styles and, 113
contour line, 197
contractions, 140
copyright, 15, 154–5
corroboration, 65
Cottingham, C., et al., 89
Courtenay, B., 215
cover letter, 67–8
Cress, J., 107
critical reading, 72
critique: annotated bibliography and, 88–9, 91
cutting-and-pasting, 40: posters and, 145, 150–1
cyclical diagrams, 150

data, aggregated, 195
databases, 21; bibliographic software and, 111; journal, 24; reports and, 73
date: citations and, 116, 119; web pages and, 30, 31
datum, map, 207
Day, R.A., 86
descriptive abstract, 68, 70
descriptive essays, 47
diagrams, 14, 16; cyclical, 150; posters and, 150; proportional pie, 195; sketch, 52
Didion, Joan, 2
discussion and conclusion section: report, 77–8
discussion lists: in note system, 133; in reference lists, 126–7
dishonesty, academic, 39–42; *see also* plagiarism
dissertations: in note system, 130; in reference list, 122; as sources, 25
distortion: maps and, 197–8
Dixon, T., 60, 255
DOI (Digital Object Identifier), 126
domain name, 30, 31
dot maps, 191–2
drafts, 4, 54–6; number of, 55
DVD: in note system, 130

Eats, Shoots & Leaves, 136, 142
Eichler, M., 142
Eisenberg, A., 65, 232
Eisner, C., and Vincinus, M., 42, 45
electronic format: submissions and, 10
electronic sources: citations of, 119; in note system, 132–3; in reference list, 124–7
ellipsis points, 37, 139
ellipsoid, map, 207
email: information dissemination and, 248–9; in note system, 133; reference lists, 126–7
Emerson, Ralph Waldo, 209
EndNote, 111
endnotes, 127, 128; *see also* note system
environmental impact statement, 42–5, 84
equipment and materials section, 76
errors: common spelling, 137; in in-text citations, 115–16
essays, 46–61; academic, 47; accuracy and, 58; argumentative, 47; definition of 47; depth of, 56; descriptive, 47; evidence and, 57–8; expository, 47; graphs and tables in, 160; headings and, 11, 12, 55; logic and, 49; persuasive, 47; plan and, 52; question and, 48–9; relevance and, 55–6; v. reports, 64; revising, 54–6; self-assessment form for, 58–60; structure of, 51–4; synthetic critical, 47; thesis statement and, 49–51; three-point, 53–4; types of, 47
et al., 118
ethics: reports and, 76, 84; research and, 42–5
evidence: essays and, 57–8
examinations, 252–70; closed-book, 253; exam-day preparation for, 259–60; guessing and, 265; help and, 259; in-term preparation for, 254–5; information on, 256–7; multiple-choice, 264–5; online, 253, 269; open-book, 253, 267–9; oral, 253, 265–7; practice and, 258; preparation for, 254–60; purposes of, 252–3; regular review and, 255; review period preparation for, 256–9; success in, 260–9; take-home, 268–9; types of, 253
examples: essays and, 57–8
executive summary, 68–71; *see also* abstracts
expository essays, 47

figures, 14, 15; citations and, 117; posters and, 154–7; reports and, 76
film: in note system, 130; in reference list, 122–3
first person: reports and, 79
flipchart, 223
fonts: posters and, 152
footnotes, 127, 128; *see also* note system
Foremski, Tom, 238–9
format: electronic, 10, 22; reference list, 13; report, 80; writing and, 8–14
fraction, representative, 202, 203
FreeAlberta.com, 32–3
frequency tables, 179
Friedman, S.F., and Steinberg, S., 17, 49, 256, 270

Game, A., and Metcalfe, A., 17
generalizations, unsupported, 57
general-reference maps, 190–1
geographic information system (GIS) software, 205, 207

Goldbort, R., 86
Google, 20–1
Google Scholar, 24
Gould, Peter, 100, 101
government reports, 24–5
grammar, 7–8
Grantham, Dewey, 92, 94
graphic scale, 202–3
graphs, 159–60; bar, 155, 163, 169–73; computer-generated, 161–3; elements of, 162; guidelines for, 160–1; line, 154, 163, 165–9; log-log, 163, 184; logarithmic axes and, 163, 182–5; manually produced, 161; pie, 163, 181–2, 195; production of, 161–3; semi-log, 163, 184; types of, 163–85
Greetham, B., 60
Gurak, L., 232
Gustavii, B., 189

handouts, presentation, 220–1
handwriting: exams and, 263–4
Harris, R.A., 19, 38, 45
Hart, C., 107
Harvard system, 110, 114–27; *see also* author–date system
Hay, I.: et al., 232–3; and Bull, J., 270; and Miller, R., 158
headings: essay, 10–12, 55; hierarchy of, 11–12; poster, 152; report, 80
Hess, G.R., et al., 144
highlighter, 255, 268
histograms, 163, 169, 173–9
Hodge, David, 78
Howitt, R., et al., 100
Huxley, T.H., 62
hypothesis: report and, 72, 76; research proposal and, 84, 85

ibid., 128
illustrative materials, 14–16; *see also* specific types
in-text system, 110; *see also* author–date system
informative abstract, 68, 69–70
inset maps, 208
instructions: exam, 260; graphs and tables in, 160
interactivity: maps and, 205; posters and, 150–1
interlibrary loan, 24
Internet: open-access movement and, 34; self-publishing and, 234, 236–7, 247–50; web and, 22
Internet sources, 20–3; in note system, 132–3; in reference list, 124–7
interviews: recording, 44
introduction: essay, 51, 53–4; oral presentation, 217; poster, 149; report, 71–2; research proposal, 84–5
inverted-pyramid style, 240–1
isobar, 197
isobaths, 197
isohyet, 197
isolines, 197, 200
isotherm, 197

Jardiel Poncela, Enrique, 1
jargon, 6, 79–80
Jenkins, A., and Pepper, D., 233
Johnson, J., 87
Johnson, Samuel, 109
Jones, M., and Parker, D., 233
journals: electronic, 125–6, 133; in note system, 129; open-access, 34; in reference list, 121; as sources, 24; style guidelines and, 113
JSTOR database, 24

Kanare, H.M., 87
keys, map, 16, 192
Kimerling, A.J., et al., 208
Kingsolver, Barbara, 1
Kneale, P.E., 17
knowledge, 19; common, 33, 38–9
Kosslyn, Stephen, 143, 160, 189
Kraus, Karl, 234
Krohn, J., 159, 189

labels: maps and, 208
laboratory research reports, 62, 64, 80, 83
language: biased/discriminatory, 134–6; media releases and, 242–3; report, 79–80
Larson, Doug, 109
legends, map, 16, 192
Lester, J.D., 60
Lethbridge, R., 151, 158
library, 23–4, 25–6
Library of Congress, 142
line graph, 163, 165–9
listservers, 248–9
literature review: report and, 72–4; research proposal and, 86
loc. cit., 128
logarithms, 183–4
logic: essays and, 49
log-log graph, 163, 184
London Underground map, 198
Lone Eagle Consulting, 234

MacEachren, A.M., 205, 208
magazines, 25; in note system, 129; in reference list, 121
maps, 14, 16, 190–208; choropleth, 195–7; citation of, 118; cognitive, 204; computer-generated, 205; concept, 258; design of, 204–8; dot, 191–2; general-reference, 190–1;

inset, 208; isoline, 197; larger-scale, 203–4; manually produced, 205; mental, 203, 204; in note system, 131; orthophoto, 201; posters and, 155; proportional dot, 192–5; purpose of, 190–1; in reference list, 123; smaller-scale, 203–4; standard elements of, 206–7; thematic, 190–1, 197; topographic, 191, 197–201; types of, 191–201
markers: as audience, 4–5; essays and, 46–7; format and, 9; scholarship and, 14; sources and, 23; supplementary materials and, 14
Marx, Groucho, 88
materials and methods section, 74–6
mathematical scaling, 193
May, C.A., 60
media: "filtered," 236; journalism-based, 236, 237–46; types of, 236–7; writing for, 234–51
media releases, 235, 236; digital environment and, 246; evidence and credibility of, 241–2; follow-up and, 246; format of, 243–5; inverted-pyramid-style, 240–1; journalism-based media and, 237–46; length of, 242–3; in note system, 130; reading level of, 242–3; in reference list, 122; targeting, 239; timing and, 243; writing, 239–44
Medical Research Council (MRC), 43
mental maps, 203, 204
methods: report and, 74–6; research proposal and, 85
Microsoft Excel, 162, 163, 180
model: oral presentation and, 223
Mohan, T., et al., 17, 87
Mondo Code, 251
Monmonier, M., 208
Montello, D.R., and Sutton, P.C., 43, 45, 87, 190
Montgomery, S.L., 63, 87, 189, 219, 233
Morris, Wright, 2
Moxley, J.M., 87
multiple-choice examinations, 264–5

National Energy Program, 32–3
National Topographic System (NTS), 201
Natural Sciences and Engineering Research Council (NSERC), 43
newsgroups: in note system, 133; in reference lists, 126–7
newspapers, 25; in note system, 130; in reference list, 122
Newton, Isaac, 18
Nick, J.M., 34
Nicol, A.A.M., and Pexman, P.M., 189
north arrow, 206
Northey, M., et al., 17, 61, 270
notes: condensing, 258; lecture, 254–5; oral presentations and, 219–20; reading, 255; substantive, 134
note system, 110, 127–34
nouns, proper, 141
numerals, 141

O'Connor, Flannery, 2
online examinations, 253, 269
op. cit., 128
open-access publishing, 34
open-book examinations, 253, 267–9
opinions: reports and, 64
oral examinations, 253, 265–7
oral presentations, 209–33; assessment form for, 211–13; audience questions and, 230–1; context and goals of, 214–15; delivery of, 227–30; delivery speed for, 220; equipment and, 215; exams and, 266; graphs and tables in, 160; material for, 215–16; notes for, 219–20; organizational framework for, 216, 217; v. posters, 144; preparation for, 214–27; recording of, 226; rehearsal for, 224–6; structure of, 216–19; time and, 214, 230; venue and, 215; visual aids for, 221–4
Orr, Fred, 255, 270
orthophoto maps, 201
outlines, essay, 51, 52
overhead projector, 223

page numbers: citations and, 115–17; report, 71
paper: using less, 10
paragraphs, 6–7
paraphrasing, 14; active, 36–7; passive, 36; sources and, 35–8
patterns: maps and, 208
PDF documents: in note system, 132; in reference list, 124–5
Pechnik, J.A., 153–4, 158, 167–8, 189
peer review, 24, 28; open access and, 34
perceptual scaling, 194
period (punctuation), 138
periodicals, 121; *see also* journals
personal communication, 119
personal experience, 57
persuasive essays, 47
Peterson, G.N., 208
photographs, 14; aerial, 123, 131; media releases and, 243, 246; posters and, 154–7; scale and, 156–7
phrases, transitional, 7
pie graph, 163, 181–2, 195
plagiarism, 5, 33–4, 35, 39–42
plates, 14
pointers, laser, 229
population: maps and, 199
population pyramid, 163, 179–80
possession: punctuation of, 140–1
posters, 143–58; assessment of, 145–8;

assessment form for, 147–8; cut-and-paste, 145, 150–1; design of, 149–57; interactive, 150–1; large-format printing and, 145; layout of, 149; organization of, 149–51; production of, 144–5, 146; text of, 151–3
PowerPoint, 210, 221, 223; drawbacks of, 222; exam preparation and, 254
précis, 91–5; assessment form for, 93–4
press release, 237, *see also* media release
primary research, 64
printing, large-format, 145
procedures section, 76
ProCite, 111
projection, map, 207
pronouns: report-writing and, 79; sexual bias and, 135
proofreading: exams and, 264
proportional dot maps, 192–5
proportional pie diagram, 195
public audiences: reasons to communicate with, 235; writing for, 234–51; *see also* audience
public speaking: career and, 210; importance of, 209–11; *see also* oral presentations
publisher: credibility and, 28
punctuation, 7–8, 136–41
Purdue OWL, 142
Purrington, C.B., 144, 158

qualitative methods, 64–5, 75
quantitative methods, 64
questions: audience, 230–1; essay, 48–9; about exams, 256; oral exam, 267; research reports and, 63–5; rhetorical, 217
quotation marks, 37, 140
quotations: block, 37–8; direct, 14, 37; page numbers and, 115, 116; sources and, 33, 35–8

Radford, M.L., et al., 28, 32, 45
ratio scale, 202–3
Reaburn, P., 73–4
readers: review and, 100–1
reading: aloud, 4, 7; critical, 72; writing and, 3
recommendations section, 78
recording: interviews and, 44; oral presentations and, 226
reference list, 12, 13, 110, 114, 119–27; bibliographic software and 111; poster, 150, 157; report, 78; research proposal, 85; review, 102–3
references, 12–13, 19, 109–34
reference styles, 112–14
RefWorks, 111
replication, 64, 75
reports, 47, 62–87; assessment form for, 81–3; v. essays, 64; expectations of, 63–5; graphs and tables in, 160; headings and, 11, 12; laboratory research, 62, 64, 80, 83; "model," 65; preliminary material of, 67–71; sections of, 66, 67; writing, 65–80
repositories, open-access, 34
representative fraction, 202, 203
research: primary, 18, 64; writing and, 3
research ethics application, 42–5, 84
research proposals, 84–6
research reports: *see* reports; laboratory research reports
results section: report, 76–7
reviews, 47, 95–107; assessment form for, 97–9; examples of, 105–7
revising: essays and, 54–6
Robson, C., 87

sample papers, 271–7; better version, 274–7; paragraphs and, 6; poor version, 271–3; purpose of, 1–2; thesis statement and, 50; titles of, 9
sampling and subjects section: report, 75–6
satellite imagery: in note system, 131–2; in reference list, 123–4
scale: cartographic, 201–2; display of, 202–3; graphic, 202–3; map, 194, 199–201, 206; map detail and, 201–4; map generalization and, 203–4; ratio, 202–3
scale factor, 201–2
scaling: mathematical, 193; perceptual, 194
scatter plots, 163–5
schedule: exams and, 257, 260–1
scholarship, 13–14
Schwegeler, R.A., and Shamoon, L.K., 61
Science Citation Index (SCI), 73
search engines, 20–1, 24, 25
"secondary hook," 241
self-assessment form: essay, 58–60; *see also* assessment forms
self-publishing, 234, 236–7, 247–50; effectiveness of, 247–9
semicolon, 139
semi-log graph, 163, 184
sentences: types of, 7
Shields, M., 61
sic, 37
sketch diagrams, 52
Slocum, T.A., et al., 191, 193–4, 208
social media: self-publishing and, 247–50
Social Science Citation Index (SSCI), 73
Social Sciences and Humanities Research Council (SSHRC), 43
software: bibliographic, 13, 73, 111; cartographic, 205; GIS, 205, 207; graphs and, 162; in note system, 131; posters and, 145; presentation, 210, 215, 221–4; in reference list, 123

sources, 12–13, 18–45; academic/non-academic, 20, 21, 22; acknowledging, 33–5; credibility of, 26–32; finding, 20–6; importance of, 19–20; incorporating, 33–9; number of, 22–3; primary, 18, 42–5; references and, 110; report, 73; secondary, 18–19; types of, 23–6; *see also* specific types
Sova, D.B., and Teitelbaum, H., 107
spam, 248
spell checker, 8
spelling, 7–8, 136; common mistakes in, 137
spider diagrams, 150
split proportional circle, 194
stereotyping, 134–6
Stettner, M., 215–16
"story hook," 241
Street, A., et al., 233
structure: essay, 51–4, 55; headings and, 10–12; oral presentation, 216–19
style guidelines, 112–13
styles: Canadian, 8; reference, 112–14
summary, 47, 88–108; annotated bibliography and, 91; assessment form for, 93–4; reviews and, 96, 99–101; writing, 91–5
superscript: notes and, 135
supplementary materials, 14–16
symbols: essays and, 71; maps and, 208; notes and, 135
synthetic critical essays, 47

table of contents: report, 71
tables, 14, 15–16, 159–60, 185–9; citations and, 117; design of, 187–9; elements of, 186–7; frequency, 179; poster, 154; report, 76
take-home examinations, 268–9
television broadcast: in note system, 130; in reference list, 122
Telg, R., 251
thematic maps, 190–1, 197
theses, 25, 62, 84; in note system, 130; in reference list, 122
thesis statement, 49–51
third person: reports and, 79
Thody, A., 79
time: exams and, 257, 260–1, 269; oral presentations and, 214, 230
timeline: research proposal, 85
title page: essay, 10; report, 67
titles: essay, 8–9; map, 206; oral presentation, 216; poster, 149; table, 15
topographic maps, 191, 197–201
Tracy, E., 270
transitional phrases, 7
Translation Bureau, 142
Truss, Lynne, 136, 142
Turabian, K., 32, 142
Twain, Mark, 219
typeface: posters and, 153

University of California at Berkeley, 45
University of California–Santa Cruz, 87
University of North Carolina, 108
University of Oregon, 20
University of Wisconsin, 87
unpublished documents, 122, 130
upper case, 141; posters and, 152
URL (Uniform Resource Locator), 30

Van Blerkom, D.L., 270
Vancouver system, 110, 112
verbal statement: map scale and, 202
videos: Internet and, 249–50; in reference list, 122
visual aids: oral presentations and, 221–4

Wainer, H., 189
Waitt, G., et al., 100
Walters, T.N., et al., 242
web, 21; Internet and, 22
web pages, 21; credibility of, 28–33; in note system, 132; in reference list, 125
websites: citations of, 119; free-access, 25; individuals and, 24; media releases and, 246; in note system, 132; in reference list, 125; review of, 95; self-publishing and, 247–50
whiteboard, 223
Wikipedia, 29, 32–3
Wilhoit, S., 108
Woodford, F.P., 66, 77
"writer's block," 52
writing: essays and, 46–61; free, 4, 52; hand-, 263–4; help with, 5–6; improving, 6–8; introduction to, 1–17; keys to, 3–6; legibility of, 9–10; length of, 8; presentation of, 8–14; reasons for, 2–3; simple, 6
writing centre, 6
writing guides, 5–6

Zittrain, J., 29